TOURISM AND MODERNITY
A Sociological Analysis

TOURISM SOCIAL SCIENCE SERIES

Series Editor: Jafar Jafari
Department of Hospitality and Tourism, University of Wisconsin-Stout,
Menomonie WI 54751, USA. Tel. (715) 232-2339; Fax. (715) 232-3200;
E-mail: jafari@uwstout.edu
Associate Editor (this volume): Graham Dann
University of Luton, UK

The books in this Tourism Social Science Series (TSSSeries) are intended to systematically and cumulatively contribute to the formation, embodiment, and advancement of knowledge in the field of tourism.

The TSSSeries' multidisciplinary framework and treatment of tourism includes application of theoretical, methodological, and substantive contributions from such fields as anthropology, business administration, ecology, economics, geography, history, hospitality, leisure, planning, political science, psychology, recreation, religion, sociology, transportation, etc., but it significantly favors state-of-the-art presentations, works featuring new directions, and especially the cross-fertilization of perspectives beyond each of these singular fields. While the development and production of this book series is fashioned after the successful model of *Annals of Tourism Research*, the TSSSeries further aspires to assure each theme a comprehensiveness possible only in book-length academic treatment. Each volume in the series is intended to deal with a particular aspect of this increasingly important subject, thus to play a definitive role in the enlarging and strengthening of the foundation of knowledge in the field of tourism, and consequently to expand its frontiers into the new research and scholarship horizons ahead.

Published and forthcoming TSSSeries titles include:

Related Elsevier journals—sample copies available on request
Annals of Tourism Research
Cornell Hotel and Restaurant Administration Quarterly
International Journal of Hospitality Management
International Journal of Intercultural Relations
Tourism Management
World Development

TOURISM AND MODERNITY
A Sociological Analysis

Ning Wang

Department of Sociology, Zhongshan University, China

2000
Pergamon
An imprint of Elsevier Science
Amsterdam – Lausanne – New York – Oxford – Shannon – Singapore – Tokyo

ELSEVIER SCIENCE Ltd
The Boulevard, Langford Lane
Kidlington, Oxford OX5 1GB, UK

First edition 2000

Library of Congress Cataloging in Publication Data
A catalog record from the Library of Congress has been applied for.

British Library Cataloguing in Publication Data
A catalogue record from the British Library has been applied for.

ISBN: 0-08-043446-0

The paper used in this publication meets the requirements of ANSI/NISO Z39.48-1992 (Permanence of Paper).
Printed in The Netherlands.

Contents

Preface

Nowadays, "to tour or not to tour?" is a question which really matters. In Western society people simply take it for granted that tourism is a necessary element of contemporary lifestyle. As a result, its significance as an industry looms large today. However, while many people recognize the importance of tourism they are usually more concerned with its contribution to the economy than its social, cultural, and political significance. Indeed, for many, tourism is still understood within a very narrow perspective and there are many other dimensions that have yet to be examined.

Tourism is a kind of social action which distances the paramount reality. It is at least partly based on the *appeal of distance* (Simmel 1990), that is, distance in time (e.g., authenticity, heritage, and primitiveness), in space (e.g., novelty, differences, physical or built environments elsewhere) and in culture or ways of life (e.g., the ethnic, the exotic, the unusual, and the abnormal). Psychologically speaking, whereas something which is too distant from one's concern may be beyond people's interest, something which is too close at hand, ordinary, and normal may lose its appeal. The appeal of distance can also be explained in sociological terms. Tourism is a religion-like ritual activity and institution (Graburn 1989, MacCannell 1973), which, through the sacralization of attractions, creates a world of hope, promise, and "salvation", or "another world" which is at a certain distance from "This World". Why do people take part in such a distancing action? At least two reasons can be given. First, they have certain problems with "this world", or with their existential conditions. Second, as a result, people question the taken-for-granted conditions of the existence in which they find themselves, and renegotiate the meaning of their lives through keeping a distance—spiritually, socially, and spatially—from everyday life and normality on a regular and annual (or semi-annual) basis.

However, within a traditional or overcoherent society, the action of distancing from normality is usually restricted and negatively sanctioned. By contrast, under modernity more people have the necessary resources to transcend the everyday world and search for experiences which are at a distance from daily experiences. Tourism thus has much to do with the conditions and consequences of modernity and offers a clue to existential problems of modern society which might otherwise remain opaque (Urry 1990a:2). In addition, tourism has become a metonym for *personalized impersonality*, a kind of social relationship characterizing modernity. Tourism is, in short, an indicator of the ambivalence of modernity. The relationship between modernity and tourism is indeed a legitimate area for study, and this is what this book intends to do.

The author's exploratory journey began in 1993 when he enrolled to study for a PhD at the University of Sheffield. He was then interested in the sociology of tourism, and felt that there was a potential "gold mine" to excavate with respect to the relationship between tourism and modernity. Part of this book (Chapters 4–6) is adapted from his doctoral thesis, another part from published articles (Chapter 2 and 3), and the rest is newly written. Most of the material was gathered in the United Kingdom and was written in English.

The author received a lot of help when writing this book. He would like to thank Jafar Jafari for his constant encouragement, advice, and help. The latter's interest to include this research theme in the Tourism Social Science Series greatly stimulated the author's enthusiasm for the venture. The author is also indebted to Graham Dann for his helpful advice, comments, and suggestions on an earlier draft. He undertook the time-consuming task of carefully reading the manuscript word by word, criticizing the text line by line, correcting errors (the errors remaining are the author's own) and polishing the English. The author would like to take this opportunity to express his gratitude to Maurice Roche, who inspired the author when he was reading for his PhD. Thanks are also due to Chris Crowther for his friendship and constant help, and to Sharon MacDonald and Chris Rojek for their helpful comments. The author is indebted to his wife, Bin, for performing all the household chores, thus enabling him to concentrate on his work. The author greatly benefited from the support given by the Department of Sociological Studies, University of Sheffield, which provided the necessary Inter-library Loan Tokens to obtain a number of important documents. Finally, thanks are extended to E. & F.N. Spon and Elsevier Science, respectively, for permission to adapt two previously published articles by the author: Chapter 2 is adapted from "Logos-modernity, Eros-modernity, and Leisure" (*Leisure Studies*, vol. 15, 1996, pp. 121–35), and Chapter 3 from "Rethinking Authenticity in Tourism Experience" (*Annals of Tourism Research*, vol. 26, 1999, pp. 349–70).

Ning Wang
17 May, 1999

Chapter 1

Introduction

It is taken for granted that the movement of wild animals is a precondition of their survival—the way to search for food and prey. However, for a long period of intellectual history, travel and movement have not been seen as essential features of the human condition. On the contrary, the sedentary state is perceived to be a characteristic of civilization. As for hordes, they are usually defined as people who have not yet been civilized and remain barbarous. The same was true of the gypsies in the past. Indeed, in civilized society the movement of populations is often associated with human tragedy: war, pestilence, flood, and drought. Thus, in Western society, subjectivity is presumably sedentary and excludes mobility (Featherstone 1995). This situation is in accord with the Western tradition of logocentrism.

Of course the movement of human beings is regarded as important, since there is constant innovation in the technologies of transportation and communication. However, in the Western sociological tradition, travel, tourism and mobility have for long been treated only as *derisive* characteristics of human beings and society, and usually as economic indicators. Although the consequences of a specific kind of spatial mobility, i.e., immigration, are well analyzed, other kinds of spatial movement, particularly tourism, have been relatively ignored. Even today the sociology of tourism is a marginal branch of sociology, and its relevance is doubted by quite a number of mainstream sociologists.

However, all of a sudden the facts speak for themselves. If tourism has constantly been growing in the post-war period, then the results of this movement were spectacular in the late 1980s and in the 1990s. "Tourism", says Crick, "represents the largest movement of human populations outside wartime" (1989:310). The masses on the move have become a spectacular landscape of consumer culture in the "global

1

village" of late modernity. The significance of tourism is, however, far from merely economic. It is also sociological.

Tourism is increasingly globalized. With touristic consumerism expanding worldwide and tourists traveling further afield, various peoples, nations, and places are becoming involved in this touristic globalization and being exposed to its positive and negative consequences. No longer can a culture or a people remain insulated. Nor can a nation be severed from international society once it has joined the enterprise of tourism, for tourism is an "international fact" (Lanfant 1980). Nowadays almost every item of culture is 'touristifiable' and can be turned into a consumer good, conveying "image", "experience", "the authentic", or "the exotic", because it has a potential audience of tourists, especially international ones.

With the arrival of the democratization, consumerization, and globalization of tourism, the latter has become integrated into the social construction of both individual and national identity. On the one hand, tourists are away from home to experience the heightened consciousness of self by searching for reference images and signs of others. On the other hand, numerous places are involved in a new or modern kind of hospitality, which is different from the traditional one, characterized by an authentic interrelationship between host and guest. This modern hospitality implies a kind of anonymous, impersonal and commercial—yet "friendly"—social relationship between hosts and strangers (a kind of relationship which is congruent with the general trend of the impersonalization of modernity). Host peoples in various places become profit-driven actors who attract tourists by turning their places of residence into spaces of spectacle, attractions, and play-grounds, a so-called "touristification" process. (Lanfant 1995b:35; Picard 1995:46). In so doing, each destination vies with others in enhancing its image of tourability (including the image of the infra- and super-structures of tourism). Thus, like social structure itself, the symbolic structure of a place appears to be significant. These touristic phenomena, being social in nature, call for serious sociological study.

The birth of modernity was in a sense signalled by tourism, which in turn was a consequence of modernity. This is a kind of spiritual resource of modernity. For instance, if the Grand Tour involved the communication of the spirit of the Renaissance, then the person of travel implied the Enlightenment. As Boorstin observes, "The travels of seventeenth century around Europe, to America, and to the Orient helped awaken men to ways of life different from their own and led to the Enlightenment" (1964:79). No small wonder, then, that the time of the Industrial Revolution in England was also the time when modern tourism came into being, as exemplified by Thomas Cook's organized tours. To risk oversimplification, its history is an alternative, although marginal, history of modernity. Rather than being merely sedentary, the modern subject is on the move (Urry 1995). As such, sociological indifference to mobility and tourism can no longer be justified.

This book attempts to reveal the importance of tourism in the formation of the modern subject and the understanding of contemporary society by studying the relationship between (late) modernity and tourism. One is not alone in choosing this as a central theme. Quite a number of pioneers have contributed, in various ways, to the sociological understanding of tourism by adopting a similar approach. This introduction will briefly review and analyze this literature and, as a result of this exercise, develop a set of ideas which can pave the way for further progress in the sociological study of tourism. The introduction covers three principal topics. First, some conceptual issues, such as the meaning of tourism, are discussed. Second, a brief review of the literature relating to the issue of modernity and tourism is undertaken. Third, the main ideas, themes, aim and the structure of the book are briefly explained.

Conceptual Arguments

One of the problems that students of tourism face is that there is no commonly accepted definition of tourist or tourism. Different definitions are used to serve different underlying purposes (Burns and Holden 1995:5; Ryan 1991; S. Smith 1988) (for a review of the literature, see Gilbert 1990; Theobald 1994a).

The words tourist and tourism did not appear before the 1500s (Leiper 1983:277). In the 1700s "tour", in the sense of "tourism", began to be used. For example Daniel Defoe used it in his book *A Tour Through the Whole Island of Britain*, which appeared in the 1720s (Leiper 1983:278). According to the *Oxford English Dictionary*, the advent of the term tourist in English was in the late 18th century, and it was used as a synonym for "traveler". Thus the meaning of "tourist" during this early period of time was neutral. Yet, while this meaning is still in currency, by the middle of the 19th century it had acquired a negative connotation, one that was diametrically opposed to the term "traveler", which had a positive meaning. Thus in the latter part of the century, when traveling abroad English people liked to consider themselves as travelers rather than tourists (Buzard 1993:1; Fussell 1980).

In defining who is a tourist, a statistical expert's definition is usually different from that of an academic. For the purpose of data relating to international arrivals in 1937, a committee of statistical experts at the League of Nations defined a tourist "as one who travels for a period of 24 hours or more in a country other than that in which he usually resides" (Quoted in Gilbert 1990:8). In 1963 the United Nations Conference on

Travel and Tourism in Rome produced the more widely accepted definition of "visitor", which was adopted in 1968 by the International Union of Official Travel Organisations (IUOTO, the predecessor of the World Tourism Organisation, WTO). It was recommended by the UN conference that the term should be divided into two categories: "*tourists*" and "*excursionists*". A tourist was defined as a person who made an overnight stay and an excursionist as one who was on a day visit:

> For statistical purposes the term "visitor" describes any person visiting a country other than that in which he has his usual place of residence, for any reason other than following an occupation remunerated from within the country visited. This definition covers:
> – *tourists*, i.e. temporary visitors staying at least twenty-four hours in the country visited and the purpose of whose journey can be classified under one of the following headings: (a) leisure (recreation, holiday, health, study, religion, and sport), (b) business, family, mission, meeting.
> –*excursionists*, i.e. temporary visitors staying less than twenty-four hours in the country visited (including travelers on cruise ships) (IUOTO 1963:14 quoted in Leiper 1979:393).

These technical and statistical definitions are characterized by behavioral and situational features, including temporal (over 24r hours), spatial (away from place of residence), and situational (not for pursuing an occupation remunerated from the place visited) elements. Therefore, such definitions provide an objective standard for internationally consistent statistics.

Such technical and statistical definitions are often, however, dismissed by some academics of tourism as too broad to capture the essential features of a tourist. They tend to define a tourist in terms of a narrower range of motivations and purposes. For example, Nash defines a tourist "as a person at leisure who also travels", and tourism as the activity of such persons (1981: 462). Similarly V. Smith stipulates that "a tourist is a temporarily leisured person who voluntarily visits a place away from home for the purpose of experiencing a change" (1989:1). Thus both travel and leisure are two necessary components of tourism (Nash 1981; Pearce 1989:1), and accordingly those who travel for non-leisure purposes (for example business) are not tourists. Such variations on what constitutes a tourist certainly satisfy academic or disciplinary interests. However, for local tourism suppliers the difference between leisure and non-leisure travelers is of little relevence. This is particularly so since there is inevitably a leisure dimension to the work of business travelers.

Cohen (1974) offers a motivational definition of "tourist", that incorporates some elements of behavioral/statistical definitions. He defines the tourist in terms of six features. The tourist is a *temporal* traveler, not a permanent traveler such as a nomad; a *voluntary* traveler, not an exile, refugee, or prisoner of war who is *forced* to travel; a traveler on a *round trip*, not an emigrant on a one-way trip; on a relatively *long journey*, not an excursion; on a *non-recurrent trip* i.e., he or she is not a commuter or a holiday-house owner; and a traveler, the purpose of whose trip is *non-instrumental*, i.e., unlike businessmen or those whose trips serve a primarily instrumental (economic, political, or religious) purpose. In sum, Cohen's definition of the tourist is as follows:

> A "tourist" is a voluntary, temporary traveler, traveling in the expectation of pleasure from the novelty and change experienced on a relatively long and non-recurrent round-trip (1974:533).

Cohen insists that the boundaries between tourist and non-tourist roles are vague and fuzzy. This vagueness partly explains the complexity and difficulty of defining who is a tourist. Unlike Nash and Smith, Cohen does not completely exclude from the universe of the tourist, business travelers and the like who travel for instrumental purposes. They may be *partial tourists*, since business travelers and pilgrims can also participate in some activities for the sake of leisure, pleasure, and recreation during their instrumental travels.

In general terms, the official, industrial, or economic definition of a tourist tends to be a technical and statistical one (broad definition). In contrast, the anthropological, sociological, or psychological definition of a tourist tends to be a conceptual or motivational one (narrow definition). Thus, the former is usually broader since the latter excludes travelers for instrumental purposes from the boundaries of the tourist.

In regard to the question of "what is tourism?", pluralism also prevails. For some, "tourism" is synonymous with the activities and impacts of the tourist (Nash 1981:462). Others give a holistic definition of the term. Thus, Leiper considers tourism as

> the system involving the discretionary travel and temporary stay of persons away from their usual place of residence for one or more nights, excepting tours made for the primary purpose of earning remuneration from points enroute. The elements of the system are tourists, generating regions, transit routes, destination regions, and a tourist industry. These five elements are arranged in spatial and functional connections. Having the characteristics of an open system, the organization of five elements operates within broader environments: physical, cultural, social, economic, political, technological with which it interacts (1979: 403–404).

Jafari also offers a holistic definition, but his is based on an epistemological approach. He defines tourism as "the study of man away from his usual habitat, of the industry which responds to his needs, and of the impacts that both he and industry have on the host's socio-cultural, economic and physical environments" (1977:6).

A similar holistic definition is put forward by Mathieson and Wall, who further distinguish "tourism" from "the study of tourism":

Tourism is the temporary movement of people to destinations outside their normal places of work and residence, the activities undertaken during their stay in those destinations, and the facilities created to cater to their needs. The study of tourism is the study of people away from their usual habitat, of the establishments which respond to the requirements of travelers, and of the impacts that they have on the economic, physical and social well-being of their hosts (1982:1).

Still others tend to define tourism from a supply-side view. Thus, while recognizing that no adequate industrial definition exists, because most definitions are based on the characteristics of the tourist, S. Smith provides a supply-side definition of tourism:

Tourism is the aggregate of all business that directly provide goods or services to facilitate business, pleasure, and leisure activities away from the home environment (1988:183).

He further divides these into two "tiers" of business. Tier 1 is composed of business and commodities which serve tourists exclusively, while tier 2 serves a mix of tourists and local residents.

Which definition of tourist and tourism will be adopted in this book? Rather than use a single definition throughout, following S. Smith's position of "intellectual tolerance and an appreciation for diversity in definitions" (1988:180), here a dual definition for different situations is employed. First, when a tourist or tourism is considered from the perspective of *demand*, an academic or motivational definition will be adopted (Cohen 1974; Nash 1981; V. Smith 1989). In this case, those traveling for *instrumental* purposes such as business, conferences, political affairs, and so on will *not* be treated as tourists. Accordingly, tourism will be regarded as the activity, experiences, characteristics, and impacts of these tourists. In other words, a tourist will be treated as a person who voluntarily travels away from home for *non*-instrumental purposes such as recreation and pleasure.

Second, when a tourist or tourism is discussed from the perspective of production, a technical, statistical, holistic, or supply-side definition of tourist and tourism will be adopted (IUOTO 1963; S. Smith 1988). That is to say, "tourist" refers to travelers who are on the move for instrumental purposes such as business and so on, as well as those who are exclusively on trips for leisure, recreation, and pleasure; and "tourism" refers not only to their activities and impacts, but also to the commodities and services supplied by businesses or the tourism industry. From this supply-side view, the distinction between instrumental and non-instrumental motives for travel is of little significance. To avoid confusion, the broad meaning of "tourism" from the perspective of tourism supply will be denoted by "tourism industry", "tourism business", "tourism economy", "tourism system", or "tourism production system" (Britton 1991).

Tourism has become a mature research topic for social scientists since the 1970s (Graburn and Jafari 1991) and sociologists have made many contributions to the study (see Allcock 1989; Cohen 1979a, 1984, 1988a; Dann and Cohen 1991; Sharpley 1994; Urry 1990a, 1991a). Recently, it has drawn wider academic attention from various disciplines, mainly because it has become one of the largest industries in the world, and "one of the quintessential features of mass consumer culture and modern life" (Britton 1991:451).

According to Cohen (1984), the first sociological account of tourism appeared in Germany in 1930 (L. von Wiese) and the first full-length work was written in German by H. J. Knebel (1960). Although similar social science writings in English were published in the 1930s (Norval 1936; Ogilvie 1933), tourism received little attention from sociologists until after World War Two. International mass tourism emerged soon after the war (partly due to the fact that the jet airplane was introduced in 1952), and there was an almost immediate academic response from students of sociology (Boorstin 1964; Dumazedier 1967; Foster 1964; Mitford 1959; Nunez 1963). However, significant progress was not made until years later (e.g., Allcock 1988, 1995; Apostoloplous, Leivadi and Yiannakis 1996; Böröcz 1996; Cohen 1972, 1974, 1979a, 1979b, 1984, 1988a, 1988b, 1995; Dann 1977, 1981, 1989, 1996a; de Kadt 1979; Greenblat and Gagnon 1983; Harrison 1992; Hitchcock, King and Parnwell 1992; Hollinshead 1992, 1996, 1997; Krippendorf 1987; Lanfant 1980, 1993; Lanfant, Allcock and Bruner 1995; MacCannell 1973, 1976, 1989, 1992; Roche 1992, 1994; Rojek 1993, 1997; Rojek and Urry 1997; Ryan 1997a; Saram 1983; Shields 1991; Turner and Ash 1975; Urry 1988, 1990a, 1990b, 1992, 1994a, 1994b, 1995; Watson and Kopachevsky 1994). Even so, and despite this trend, the sociology of tourism "is still very much in its infancy" (Dann and Cohen 1991:158).

For a subject to come to the point of maturity at least one of the following conditions must be satisfied. First, there should be "legitimate territories" within which a field or discipline is located. A discipline has its own "sovereignty" to which other disciplines cannot easily lay claim. Thus, for example, political science, economics, sociology, psychology, and geography all have their own domains. Although a cross-disciplinary perspective may become necessary in studying more and more social phenomena, each must nevertheless retain its own identity. Second, distinctive approaches, perspectives, or methods should be used to study the subject matter of "legitimate territories".

The difficulty for the sociology of tourism is that it has no monopolized "legitimate territory". In this regard, Cohen (1984) has identified four principal areas for the sociological study of tourism: the tourist, relations between tourists and locals, the structure and functioning of the tourism system, and the consequences of tourism. One cannot argue with such a classification. However, these four domains are, of course, not the "legit-imate territories" that are monopolized by the sociology of tourism. Other disciplines can also lay claim to these areas. For example "the tourist—his motivations, attitudes, reactions, and roles" (Cohen 1984:373) has also been studied in depth by psychologists (Iso-Ahola 1983; Pearce 1982, 1988; Ross 1994). The relationship between tourists and locals seems to be principally the domain of anthropologists (Nash 1981; V. Smith 1977, 1989). The structure of the tourism system has been examined by tourism studies in general (Cooper, Fletcher, Gilbert and Wanhill 1993) and social geography in particular (Britton 1991; Shaw and Williams 1994). The socioeconomic and cultural impacts of tourism have also been investigated by holistic tourism studies in general, and social geography in particular (Mathieson and Wall 1982; D. Pearce 1989). Tourism is essentially a multidisciplinary study. Hence, there are no clear-cut territories reserved *exclusively* for the sociology of tourism, although some areas clearly call for more sociological treatment than others.

Since it is difficult for the sociology of tourism to claim its own "sover-eignty" over a monopolized "territory", it can only claim legitimacy through its own distinctive approaches, perspectives, and methods. In this respect, the sociology of tourism can be justified as a legitimate area of study, since it offers distinctive sociological approaches and per-spectives to tourism which other disciplines cannot.

Interestingly, even if one can intuitively tell a sociological approach from that of another discipline, such as psychology or economics, it is still difficult to define clearly what a distinctive sociological approach is. In reality, sociology is quite controversial in terms of its own approaches and perspectives. With regard to tourism, there is no single approach but rather numerous sociological ones (Dann and Cohen 1991). These approaches include the Weberian, or tourism as meaningful action and motivation (Dann 1977, 1981); the Durkheimian, or tourism as ritual and myth (Graburn 1989; MacCannell 1973, 1976; Selwyn 1996a); the Marxian, or tourism as false consciousness and ideology (Thurot and Thurot 1983); the structural-functional, or tourism as social therapy (Krippendorf 1987); the structural-conflictual, or tourism as the conflict of interests between the Core and Periphery (Turner and Ash 1975); the symbolic interactionist, or tourism as communication of identity and as symbolic display of status (Brown 1992; Dann 1989); the phenomenolo-gical, or tourism as experiences (Cohen 1979b; Ryan 1997a); the feminist, or tourism as gender inequality (Kinnaird and Hall 1994); and the post-structuralist, or tourism as sign, discourse, and representation (Culler 1981; Dann 1996a; Lash and Urry 1994). While all these approaches are sociological, there must still be a distinctive common identity for all of them. This identity, it is argued here, lies in a more *holistic* treatment of the subject matter in comparison with other disciplinary approaches. Economics, politics, geography, and psychology are all characterized by abstracting their own subject matters, such as the economy, polity, spaces, and psychological phenomena, from the rest of social reality, as if they were independent. By contrast, a sociological approach treats any phenomenon (such as values, activities, and social processes) in terms of human interaction, or in relation to other social phenomena, the wider context of social trends, social structures, or social demography. The micro-sociological approach, as a legitimate approach, does not fail to consider these wider contexts, although its focus is on micro-situations.

Within the community of sociologists, how to treat tourism as a legitimate area of study is also controversial. As noted previously, for a number of mainstream sociologists, tourism is a trivial pursuit and thus not worthy of serious academic effort. As a result, this subject is still regarded as a marginal branch of applied sociology. Even for those who acknowledge tourism as a legitimate academic area, there are problems of how to locate tourism on the map of sociological exploration. In charting tourism, sociologists usually examine it in three ways (Dann and Cohen 1991). First, tourism is treated as a *sub*set of leisure by the sociology of *leisure* (Dumazedier 1967; Krippendorf 1987; Rojek 1993). Second, it is regarded as a specific kind of migration, such as seasonal leisure migration (Böröcz 1996; Vukonić 1996). Third, it is legitimized as the subject matter of the sociology of tourism in its own right, an approach which stresses the *travel* dimension of tourism (Cohen 1972, 1979a, 1984, 1988a; Graburn 1983a, 1989; MacCannell 1976; Urry 1991a, 1995). Whichever approach is adopted, a number of concepts are employed to characterize tourism sociologically, such as escape (Cohen and Taylor 1992; Dumazedier 1967; Rojek 1993), social therapy (Krippendorf 1987), authenticity-seeking (MacCannell 1973, 1976), quasi-pilgrimage or ritual (MacCannell 1973; Graburn 1983a, 1989), play (Cohen 1985; Mergen 1986), the core and the periphery (Britton 1982, 1991; Turner and Ash 1975), strangerhood (Greenblat and Gagnon 1983; Böröcz 1996), consumerism (Watson and Kopachevsky 1994), leisure migration (Böröcz 1996), and discourse (Dann 1996a).

One fundamental approach that sociologists apply to tourism, whether consciously or not, is what can be called the "*contextualism of modernity*". It can be argued that the study of the relationships between modernity and tourism is a central, if not the whole, task of the sociology of tourism. Indeed it was laid down by the pioneers of the sociology of tourism: Boorstin (1964), Dumazedier (1967), MacCannell (1973, 1976), Cohen (1972), and Dann (1977, 1981). Boorstin's (1964) cynical critique of mass tourism as a depthless "pseudo-event" in America may thus be understood as the first attempt to explore the relationship between tourism and modern society. Dumazedier (1967) treated the phenomenon of the "mass on the move" in terms of urbanized and industrial society. He revealed tourism as an escape from the alienation arising from an urban way of life. Thus, his exploration was another attempt to study tourism in terms of the wider context of modernity. However, MacCannell (1973, 1976) was the first writer *clearly* to relate tourism to the sociology of modernity. In response to Boorstin's hostile attitude towards mass tourism, MacCannell treated tourism as a ritual celebration of the differentiation and wholeness of modernity, and also as a quasi-pilgrimage—quest for the authenticity and meaning which were lacking in the home society but which existed in other places and other cultures. Rather than dismissing tourism, he regarded it as an integral element of modern life. He saw the tourist as one of the best models for the modern individual. Tourism for him was thus a cultural phenomenon that mirrored the structure and contradiction of modernity. Cohen (1972), although in a different way, viewed tourism as an essentially modern experience and, in so doing, confirmed the possibility of studying this subject in the context of modernity. This contextual approach was soon adopted by many of their successors. For example Dann's (1977, 1981) studies of tourist motivation linked the latter to the context of industrial modernity. He argued that tourists travel because they want both to escape the "anomie" (normlessness, meaninglessness, and isolation) of modern life, and to compensate for the dissatisfying aspects of everyday life, such as relative status-deprivation, with the "ego-enhancement" of tourism. Based in part on Foucault's approach, Urry (1990a) has introduced another influential paradigm for the sociology of tourism—the "tourist gaze", around which various power relations involving consumption and production are analyzed. Tourism has also been explicitly linked to more all-encompassing paradigms—both modernity and postmodernity (1990a, 1995). Rojek (1993) has investigated how tourism is socially organized as "ways of escape" under the condition of modernity. Böröcz (1996) has similarly explored how modernity is related to "travel capitalism", and how the uneven development of tourism is determined by differences in the degree of industrialization and modernization among some Western and Eastern European countries.

The sociology of leisure and the anthropology of tourism have employed escapism and compensation in their studies of tourists and their experiences. Thus, tourism as a form of leisure is regarded as a (temporary) escape from the alienation, monotony, etc. of everyday life (Cohen and Taylor 1992; Krippendorf 1987). As far as compensation is concerned, tourism is seen as a "repayment" for the limits of everyday life, a ritual inversion or reversion of "ordinary life" or a ritual intensification of non-ordinary experience (Gottlieb 1982; Graburn 1983a; Lett 1983). These approaches, particularly when they are combined, are still useful and valid and have been successfully incorporated into sociological studies of motivation and experiences. However, their deeper significance has been revealed only after they have been linked to the broader approach of the "contextualism of modernity". This orientation was, as mentioned above, introduced to the sociology of tourism by its pioneers more than twenty years ago. It is within this context that the concepts of escape and compensation have obtained deeper sociological significance. Thus, tourism, which had hitherto been considered by many as a superficial and trivial topic, began to gain "a deeper structural significance" (Cohen 1979a:22). Indeed, the continuous and stubborn growth of tourism during the post-war period has promoted sociologists to ask deeper questions in regard to this phenomenon. Why do people travel? Is it enough to treat tourism simply as a quest for pleasure, or as the natural outcome of an increase in discretionary time and income? Besides an improvement in living standards, is there anything else that is responsible for the emergence and growth of tourism? Does not tourism indicate that there may be something wrong with the existential condition of modernity? Is not tourism an opiate inasmuch as modernity uses it to seduce people to its own exciting but problematic order? All of these important issues deserve further debate and discussion.

This book elaborates on the relationship between tourism and modernity. It studies tourism within the context of modernity which has developed in the tradition of sociology. Tourism is thus no longer simply regarded as a universal and homogeneous phenomenon. Rather, as many pioneers have pointed out, it is essentially a contemporary phenomenon and thus needs to be analyzed in terms of the larger context of modernity. The justification for the present work is that, although this approach has been developed in the classic sociological writings on tourism, the academic "fruits" of this particular tree (the contextualism of modernity) are still relatively deficient. Much more can and should be achieved. Therefore, the aim of this book in choosing such an approach is to elaborate upon the theme and demonstrate its importance to the sociological theory of tourism.

Clearly, sociology cannot tell the whole story. That is the task of more than one discipline (Dann and Cohen 1991:167; Graburn and Jafari 1991; Przeclawski 1993). Therefore, this book does not pretend to reveal all aspcts and issues of tourism. It insists on a *sociological* approach since this is indispensable in telling part of the story, but some other disciplinary studies, for example in the fields of geography and anthropology, are increasingly overlapping with sociological studies of tourism (Shaw and Williams 1994; V. Smith 1989). Further, this book argues that there is no *single* sociology of tourism, but a number of sociologies of tourism (Dann and Cohen 1991:167). The contextual explanation in terms of modernity is just one—albeit an important one—of sociologies of tourism.

Tourism and the Ambivalence of Modernity

One commonsense view is that tourism is universal and has existed throughout history, for human beings have a certain innate need for recreational travel. However, several sociologists argue that tourism is essentially a modern phenomenon (Böröcz 1996; Cohen 1972:165, 1995:12; Dumazedier 1967; MacCannell 1976; Saram 1983:99; Urry 1990a), and Böröcz maintains that "the notion of tourism as a transhistorical constant of human life is not very useful" (1996:49). Therefore, for many sociologists, tourism can be better understood within the context of modernity.

One could say that tourist demands and motivations are mostly biological and psychological (for example north-west European (holiday-makers) travel to enjoy favorable weather in southern Europe), so how can they have anything to do with modernity? However, biological impulses or psychological factors are intertwined with social environments, as well as structural and cultural conditions, indicating that tourist motivations and demands are not purely biological or psychological, but also sociological. Modernity has established its norms and mechanisms to regulate, by either constraining or releasing impulses and needs (Elias 1978, 1982; Elias and Dunning 1986). To a certain extent, biological instincts or spontaneous drives are negatively sanctioned by society and culture, that is, they are required to be constrained and subdued (as in the realm of work and production). On the other hand, they are positively sanctioned by society and culture and hence allowed to be released and gratified, but within a liminal zone, such as a paid holiday at a resort (Elias and Dunning 1986; Shields 1991; Urry 1990a; Wang 1996). Therefore, as Krippendorf puts it, "man was not born a tourist. . . . What drives millions of people from their home today is not so much an innate need to travel. . . . The travel needs of the modern age have been largely created by society and shaped by everyday life" (1987:xiv). To understand the formation of tourist motivations, consciousness and consumption, it is necessary to appreciate why and how people under the condition of modernity, are transformed into tourists.

The formation of tourist motivation is not merely an issue of bio- or psychogenesis at the level of the individual, but also a question of sociogenesis at the levels of society and culture. To put it another way, the formation of this motivation involves the development of certain modern values "about health, freedom, nature, and self improvement" (Graburn 1983a:15), which are closely related to modernity and are also "social facts" in a Durkheimian sense (Lanfant 1993) or a "total social phenomenon" (Lanfant 1995a:2, 1995b:26). These values act as the cultural sanctions of people's biological and psychological impulses and desires, and shape an individual's consciousness of and attitude towards tourism.

Thus, in certain cases, people may have a holiday less on account of their innate needs and more due to the pressure of the norms created by society, that is, the pressures of a possible "stigma of absence" linked to the disabled body and career failure. As Graburn states:

> Within the framework of tourism, normal adults travel and those who do not are disadvantaged. By contrast, able-bodied adults who do not work when living at home are also in a taboo category among contemporary Western people (1989:23).

Therefore, with respect to tourist motivation, in contrast to a psychological perspective that focuses on experiential and psychological factors sociology concentrates on cultural values and the social mechanisms that help shape these values. Rather than exploring psychological motivations, sociology studies both cultural conditions (such as values and social consciousness) and structural conditions (such as enabling conditions) that are responsible for the sociogenesis of tourism. There is, therefore, justification for a sociological study of tourist motivation on the basis of the wider relations between culture and modernity.

Nevertheless, some say that premodern persons occasionally traveled for pleasure (e.g., pleasure travel in ancient Rome). As Nash claims, "I believe that there is ... some form of tourism at all levels of human culture. To satisfy some of our critics we may have to call it 'prototourism,' but it is tourism nevertheless" (1981:463). How can tourism exclusively be linked to modernity? Indeed, certain premodern people traveled, and some of this travel might have had certain intrinsic characteristics similar to those of modern tourism (such as pleasure). However, there are several differences between premodern travel and modern tourism. First, premodern and modern people have different orientations, attitudes, and conceptions of recreational travel. In premodern society, tourism was not a socially and culturally accepted lifestyle, phenomenon or leisure activity. By contrast, under modernity, especially late modernity, tourism has become widely accepted as part of life, and for many it has become a deeply rooted habit. Furthermore, this orientation towards tourism has been increasingly globalized under late modernity. This has transformed tourism into a virtual "necessity": whereas in the past tourism was a luxury, available only to élite groups, in modernity and late modernity, tourism is for mass consumption. In association with the view that freedom of movement is a basic civil (or political) right, in line with freedom of association and communication, tourism is nowadays seen as a form of welfare, a "social right", "an important indicator of social well-being" (Haukeland 1990:179). With the arrival of the "democratization of travel" (Urry 1990a, 1992) in the West in the post-war period, "to travel or not to travel" has consequently become a social question. Lack of the opportunity to travel is treated as a sign of social deprivation. In this sense there is a touristic right in modernity (Urry 1995), and this right is a "total social fact" (Lanfant 1993:77) which is connected with broader aspects of participatory citizenship in modern societies.

Second, another major difference between modern tourism and its predecessor relates to the necessary social conditions. In today's society, there is massive social organization (Urry 1990a) or a "tourism production system" (Britton 1991). As Turner and Ash point out, nowadays, the unifying factor "is not their [tourists'] motives or attitudes ... but the existence of a coherent industry which strives to recognize, stimulate and serve the travel needs of all of them" (1975:14). The commodification of tourism, as part of overall capitalist commodification, is a form of social organization and the production of experiences is based on the logic of capitalistic commodity production (Britton 1991; Watson and Kopachevsky 1994). Thus, if premodern travel was related to risk, hardship, and travail, then today's tourism is a consumer good, a commodity characterized by safety, ease, and comfort and involving complicated social relationships that are integrated within tourism.

Third, while premodern travel was an occasional event, modern tourism is a mass phenomenon, an *institution,*—institutionalized leisure travel concentrated into a number of consecutive days, usually in the form of holidays with pay. Thus, although modern tourism can be described as a quasi-pilgrimage that "humans use to embellish and add meaning to their lives" (Graburn 1989:22), it is above all an institutionalized leisure and consumer activity characterized by pleasure-seeking, the "tourist gaze" (Urry 1990a), seasonal leisure migration (Böröcz 1996; Crick 1989:327; Vukonic 1996:31–3), the ritual inversion of everyday roles and responsibilities (Crick 1989:332; Graburn 1989; Gottlieb 1982; Leiper 1983), evasion (Saram 1983:93), escape (Rojek 1993), and social therapy (Krippendorf 1987). Although tourism in modern society is a marginal activity, a deviation from everyday roles, it is nonetheless institutionalized and is often culturally and socially constructed as an annual ritual, functioning as the "lubricating oil of pleasure" that keeps daily life going. As Krippendorf points out, "People travel so that they may be confirmed in the belief that home is not so bad after all, indeed that it is perhaps the best of all. They travel in order to return" (1987:xvi). Touristic deviation from everyday life therefore serves to renew the meaning of home and to reinforce order in daily routines.

The history of tourism in Western modernity, either in terms of tourism as a form of leisure travel, or as a specific commodity production system, coincides roughly with the history of modernity. Thus, the relationship between tourism and modernity is worth examining in depth. The question of how tourism is causally related to modernity is interpreted in various ways within the literature. According to one body of opinion, tourism originateds from modern people's reaction against and resistance to the dark side of modernity. Accordingly, tourism is treated as an *escape* from the alienation of modernity (Cohen and Taylor 1992; Rojek 1993). In other words, tourism is a mirror of disenchantment with modernity; the sociogenesis of tourism is described in metaphorical terms as the "push of modernity". By contrast, another body of opinion has attempted to demonstrate that tourism is in fact a "false" necessity, and that the demand for tourism is the result of manipulation, seduction, and control by the tourism production system (a sector of capitalist commodity production) (Britton 1991; Watson and Kopachevsky 1994). Hence, those holding this view explain the social origin of tourism mainly in terms of the "pull of modernity".

Both positions contain elements of truth. Yet they are partial if they exclude or oppose each other. In fact, the "push" factors and "pull" factors are two sides of the same coin (Dann 1981). They indicate the same *structural ambivalence* of modernity from different aspects, and it is this which underpins the sociogenesis of modern tourism. One can say that modern tourism is a cultural celebration of modernity (such as the improvement of living standards, and increased discretionary time and disposable income), appearing as tourism-related consumer culture. One can also say that it is a cultural critique and negation of modernity (such as alienation, homelessness, stress, monotony, and urban environmental deterioration), exhibited as an escape and a desire to "get away from it all" (home and daily responsibilities). Tourism can be both. It is an expression of both "love" and "hate" in response to the existential condition of modernity. To be away from home implies returning home again. Being "home and away" is a persistent touristic dialectic, reflecting the deep structural ambivalence of modernity.

In general terms, "modernity" refers to the period since the Renaissance and is thus associated with the *replacement of traditional society* (premodernity). More specifically, "modernity" refers to a new social order that has arisen during the last two or three centuries, a social order that first appeared in the West and then spread to the rest of the world. It comprises an institutional order (capitalism, industrialism, surveillance, and the monopoly of violence by the nation-state) (Giddens 1990), an intellectual order (science and technology, de-enchantment) (Weber 1978), a temporal order (schedulization, synchronization, routinization, or accelerating tempo and rhythm) (Simmel 1990), and a sociospatial order (urbanization, globalization, abstractization of space, etc.) (Lefebvre 1991). All these dimensions are in reality intertwined with one another. At the heart of the institutional, intellectual, temporal and spatial orders of modernity is "rationalization" (Weber 1978), a process whereby traditional customs give way to contemporary ways of doing things. Many authors argue that postmodernity replaced modernity during the last quarter of the twentieth century (Harvey 1990). However, this idea is still controversial. This book does not deny the structural changes suggested by the term "postmodernity". However, since these so-called postmodern changes have not transcended rationalization, this book, following Giddens (1990), disagrees with the view that postmodernity has already replaced modernity. On the contrary, postmodernity is viewed as a new form of the same order (rationalization). Classical modernity and so-called postmodernity are two different forms of the modern order. They are two analytical devices used to characterize different phenomena within the same contemporary society. Therefore, it may be better to treat so-called postmodernity as late modernity, in reference to the forms of social organization characterizing advanced society during the last quarter of the twentieth century. Therefore, when the term "modernity" is used, it may refer to early or late modernity, or both, but, more often than not it refers to late modernity.

The term "ambivalence" was introduced by Eugen Bleuler earlier in the 20th century, since then it has been both employed and explored mainly by psychologists. The "concept of ambivalence in psychology refers to the experienced tendency of individuals to be pulled in psychologically opposed directions, as love and hate for the same person, acceptance and rejection, affirmation and denial" (Merton 1976:6). However, according to Merton, ambivalence is not only a psychological concept, it is also a sociological one. Whereas the psychology of ambivalence focuses on personality, its sociological equivalent highlights "how and to what extent ambivalence comes to be built into the very structure of social relations" (1976:5). The latter "refers to the social structure, not to the personality" (1976:6). Thus, "sociological ambivalence is one major source of psychological ambivalence" (1976:7). The analysis of the sociological ambivalence can be traced back to Freud (1963), who demonstrated how civilization is essentially ambivalent, and that all cultural life includes an imprint of the ambivalence of civilization. Following Freud, Elias (1978, 1982) empirically demonstrated how the "ambivalence of interests" in social relations led to the formation of the absolutist state and associated forms of culture, such as manners. Likewise, as a cultural phenomenon, tourism also has roots in the structural ambivalence of modernity. This is a central theme of this book.

Modern tourism involves the interaction between consumers and tourism-oriented capital. On the one hand, the presence of the tourist as a consumer implies that there is sufficient, public or private investment in the facilities and infrastructure which are necessary for consumption, such as attractions, transport, and accommodation. On the other hand, capital requires a mature tourist market, namely a sufficient number of people with the desire to consume tourism as a commodity. Justice cannot be done to either of them if only one is acknowledged at the expense of the other, although only one of them can be the analytical focus of a given piece of research due to the social division of academic labor.

The emergence of the tourist has to do with the *enabling* conditions of modernity, metaphorically the so-called "love" side of the ambivalence of modernity (or the "pull" of modernity). As Marx and Engels claim in *The Communist Manifesto*, one of the most striking characteristics of capitalism is unprecedented productivity. Greater productivity not only necessitates but also facilitates a faster flow of commodities, people, and capital, which prepares the social conditions for the constant advancement of transportation and communication, and each technological revolution in transport and communication stimulates the further enhancement of productivity. Transport, communication, and travel are therefore integral elements of the system of capitalist commodity production and circulation. Thus, for the sake of commodity production and exchange, it is necessary for a society to devote a certain amount of capital to the facilities and infrastructure that are indispensable to transportation and communication, the basis upon which tourism can develop. That is why the first industrialized countries were the first tourist destinations. Thus, Britain, for example, was not only an early industrialized country, but also one of the first developed tourist destinations in the world, enjoying high visitation rates; as Böröcz notes:

> The penetration of leisure migration presupposes the availability of the services and infrastructure used for commercialized travel. That requires a certain level of surplus in the society at the destination, so that labor and infrastructural resources can be devoted to the service of foreigners and the transformation of social structures into ones capable of and willing to accommodate a primary commercial flow of strangers (1996:28).

The development of a tourism industry, therefore, is closely related to infrastructural development and the ability to accommodate a large volume of leisure and commercial travelers. However, adequate infrastructure is a necessary but not sufficient condition for the sociogenesis of tourism. The emergence of tourism also entails a social condition, namely, entrepreneurs who are willing to devote their capital to the commercialzation of tourism. In this respect the advent of Thomas Cook's tours signaled the beginning of the commercialization of tourism, a form of social organization based on technological advances (Urry 1995), which are integral to capitalist commodity production and exchange in general. For those Third World countries that have sufficient tourist resources but not the capital to devote to infrastructure and facilities, the inflow of foreign capital is unavoidable. This capital integrates these countries into the global system of the capitalist production of tourism. Nowadays, almost all countries are within the reach of tourist-sending societies and are accordingly involved in the inflow of capital from commercial intervention by the latter.

From the demand side, the emergence of tourism also has to do with certain levels of productivity (Nash 1989:39, 40, 41). Higher productivity creates spatial mobility (improved means of transport and travel) on the one hand, and psychological mobility (the desire to travel) on the other (Nash 1989). Indeed, with an increase in productivity, together with an enhanced and just distribution system, the number of people who have a surplus over "bare subsistence" grows (Pimlott 1976). They are able to transfer "a certain amount of surplus value to wages spent on such types of nonessential consumption as leisure travel" (Böröcz 1996:28). Moreover, higher productivity also indicates an increase in leisure time. "Less and less time is spent by most individuals in the world of productive labor of any kind. Consequently, more and more time is spent in private life" (Berger, Berger and Kellner 1973:171). As long as people are willing, or cultures encourage, they are able to spend part of their disposable leisure time on holidays.

"To travel or not to travel?" is a question involving the identity of modern citizenship. "We" travel for pleasure and fun because "we" are moderns. "They" don't travel because "they" are socially and economically constrained from doing so, and hence are still outside the modern lifestyle. Therefore, tourism, especially mass tourism, is an indicator of the affluence brought about by modernity and its associated lifestyles. The rate of national participation in tourism becomes one of the indicators of a demarcation between the traditional and the modern. This symbolic aspect is exploited by the cultural branch of tourism, i.e., in tourism advertisements. Touristic consumer culture also takes symbolic meanings of tourism as a concern. In this sense, people's leisure travels are a cultural celebration of the "love" side of the ambivalence of modernity.

However, tourism is also a popular expression of disenchantment with the "hatred" side of the ambivalence of modernity. Urry (1990a:2) argues that tourism can be treated as similar to deviance, for it is also a deviation from "normal society". Therefore, the study of tourism helps clarify what is wrong with this "normal society", which might otherwise remain opaque. Thus, tourism can be regarded as a kind of responsive activity that implies a critique of the dark side of modernity. Alienation or inauthenticity, the degradation in the environment, stress, monotony, and homogenization are all expressions of the "hatred" side of the ambivalence of modernity. People's loathing of the "evils" of modernity can be either verbal or non-verbal. Tourism is a non-verbal critique of these evils, for people's disenchantment with the dark side of modernity is deeply—albeit sometimes unconsciously—rooted in their motivations for tourism.

Tourists' disenchantment with the dark side of modernity is unconsciously expressed in their need to perform a kind of role and pursue a type of lifestyle that is contrary to their normal and daily lives (Gottlieb

1982; Lett 1983; Rojek 1997:58). People do not simply take a holiday. Their choice of holiday, unconsciously, or consciously, reflects their desire to change the order of 'everydayness', where the dark, as well as the positive side of modernity is embedded. They change this order by escaping (Rojek 1993), by engaging in "extraordinary" experiences (Urry 1990a), by searching for authenticity (MacCannell 1976), by experiencing novelty (Cohen 1972), by "anomie-avoidance" or "ego-enhancement" (Dann 1977), or by a quest for simplicity and the exotic (Turner and Ash 1975). All these motives help to boost seasonal leisure migration (Böröcz 1996).

All migration indicates disenchantment with something in the home society, including "hard" evils such as poverty, suffering, and political persecution, and "soft" evils such as the monotony, routinization, stress, and alienation that are closely intertwined with the goods of modernity, such as higher living standards. If modernity has, in a material sense, broadly eliminated "hard evils", then it is destined to relate to its "soft evils". The latter, mostly spiritual, constitute the "hatred" side of the ambivalence of modernity. People from Third World countries engage in permanent, or at least long-term, migration to developed countries in order to escape the "hard evils" of the home society. By contrast, people from advanced countries migrate to the "pleasure periphery" of less developed countries (Turner and Ash 1975) in order *temporarily* to escape the "soft evils" of the home society. They get away in order to return with renewed meanings of home. Boorstin summarizes the contrast between these two categories of migrant as follows:

> Men who move because they are starved or frightened or oppressed expect to be safer, better fed, and more free in the new place. Men who live in a secure, rich, and decent society travel to escape boredom, to elude the familiar, and to discover the exotic (1964:78).

Tourism involves a temporary change of the *status quo*. However, it ends up as protection and reproduction of the *status quo*. It is, therefore, conservative in effect. MacCannell regards tourism and revolution as "the two poles of modern consciousness—a willingness to accept, even venerate, things as they are on the one hand, a desire to transform things on the other" (1976:3). Yet one could argue that tourism is also a willingness to change, to alter the present order, to destroy current prohibitions and norms at least temporarily; it thus shares with revolution a common feature—it changes the present order. However, the difference between the two is still obvious; revolution wants to alter things permanently (whether or not it will be successful is another matter), whereas tourism changes the present order only temporarily, fantastically, and illusively. Tourism modifies reality by means of escape into qualitatively different spaces. Thus, it is a way of *avoiding* the present order. Therefore, if revolution is a radical form of change, then tourism is essentially a conservative form of change, serving to consolidate the everyday order at home. Thus, unlike revolution, or what Marx calls a "weapon's critique" of capitalism, tourism is only a kind of "euphemistic critique" of modernity that moderates the disappointing aspects of modernity. In paraphrasing MacCannell, Van den Abbeele writes:

> Thinking he is engaging only in his own pleasure, the tourist is unconsciously contributing to a "strong society". Tourism is thus an institutional practice which assures the tourist's allegiance to the state through an activity which discreetly effaces whatever grievances, discontent or "alienation" that the tourist might have felt in regards to society. The tourist enslaves himself at the very moment he believes himself to have attained the greatest liberty. Tourism, to paraphrase Marx, is the opiate of the (modern) masses (1980:5).

Thus, tourism, like religion, functions as the opiate of the masses and helps reproduce the *status quo*. Politically speaking, it acts in complicity with the state in the reproduction of the social order. Tourism is neither simply a freedom, nor simply a result of manipulation by the tourism industry. It is, rather, a responsive action to the ambivalence of the existential conditions of modernity, but it ends up helping to reproduce these existential conditions.

Furthermore, although tourism acts, at least partly, as a cultural "rebellion" against that capitalist commoditization which has destroyed authentic human relationships, it itself comes into being with the help of capitalist commoditization. Thus, tourism appears as a response to the ambivalence of modernity, but finishes up as ambivalence itself, namely both "love" of and "hatred" of the modern commoditization of travel experiences. Indeed, mass tourism is made possible by the tourism industry, i.e., the commoditization of tourism. However, this leads to homogeneous, standardized, and inauthentic tourist experiences (MacCannell 1973, 1976), which increasingly causes dissatisfaction among tourists.

This ambivalence is experienced not only by tourists but also by tourist destinations. Unlike goods that are used by consumers in their homes after being purchased, most of the components of tourism products are intangible, and are consumed at the point of destination, simultaneously with the period of travel. This greatly increases the amount of personal contact between hosts and guests. As many of the destinations that are integrated into the network of Western tourist consumerism are economically weak and many locals are forced to perform marginal jobs, such as service jobs, certain tensions arise from these contacts. The economy of tourism does provide locals with jobs and other economic benefits, metaphorically the so-called "love" side of the coin, but there are also "hatred" components. The relationship between local and tourist is asymmetrical. On the one side is the tourist who is engaging in leisure, play, and recreation. On the other side is the local employee who works and serves (Nash 1989:45). Tourism thus involves two categories of people:

> those who serve and those who are served. As such, it is not hard to see how feelings of superiority and inferiority develop in tourism relationships, and why it is the locals who more often than not must adapt to tourists' wishes, demands, and values, and not the other way around (Watson and Kopachevsky 1994:653).

A shopkeeper in the US Virgin Islands has voiced a similar complaint: "The only trouble is that the tourist is here for fun, for a party. We're here all the time and nobody can be in a happy-happy party mood all the time" (O'Neil 1972:7, quoted in Britton 1979:324). In Third World destinations such "hatred" components of the ambivalence of tourism may sometimes be more intense than, or even fuse with, sentiments of nationalism, which may be leveled against white tourists. Tourists are also sometimes taken by terrorists as hostages in order to place pressure upon a government for a political goal. Thus, tourism may be either a "sunny" enterprise or clouded with malaise or potential danger. From a deeper structural perspective, the ambivalence of the tourism economy is one manifestation of the asymmetrical world-system in which the core (the major transnational companies in affluent countries from whence tourists originate) dominates the periphery (the less developed tourist-receiving areas) (Turner and Ash 1975). Under such international conditions, the economy of tourism in the Third World may enter into a "dependency syndrome" (Dann and Cohen 1991:162). In short, the ambivalence of the tourism economy is the embodiment of the structural ambivalence of globalizing modernity.

Finally, it needs to be pointed out that, as modern tourism originated in the West, when modernity is referred to here, it is mainly Western modernity which is the focus. However, since there are some fundamental structural similarities (i.e., market economy, industrialization, bureaucracy) as well as differences between Western and Eastern modernity (especially Japanese modernity), the concept of modernity should be understood as incorporating the contemporary experiences and roles of some Eastern societies, such as Japan and Singapore. In addition, while modernity as a whole is closely connected with the development of tourism, it is *late* modernity that is particularly so associated, and it is this link that is the current focus. Late modernity includes the condition of globalization. For present purposes, "post-modernity" is treated as a dimension of late modernity.

Aim and the Structure

The aim of this book is to theorize tourism sociologically from various perspectives, some of which are outside the domain of tourism study. The underlying methodological principle is that of falsification (Popper). According to Popper (1969), science begins with the formulation of hypotheses which are then subjected to empirical falsification, rather than starting with the induction of empirical facts, upon which theories are subsequently formed and verified. Thorough verification of a theory is impossible, but it is open to falsification. What distinguishes metaphysics from science is not whether or not it can be verified, but rather whether it dares face falsification. Of course, in reality every hypothesis is formulated on the basis of certain empirical facts. However, the relationships between hypotheses and empirical facts are not linear but dialectical, involving constant interplay with each other. Therefore, theory is always tentative and suggestive, facing constant falsification by endless empirical evidence. Accordingly, there is no intention to claim that the arguments raised in this book are conclusive. It is believed that they will only be completed when they evoke critiques.

It can be argued that one shortcoming of tourism studies is overempiricism. Although the numerous empirical studies on tourism from various perspectives have certainly contributed to an increase in knowledge, too often they are satisfied merely to individual cases without any attempt to relate these to theory.

Theorizing can be carried out from different perspectives: economic, political, psychological, geographical, and sociological. As sociological perspectives are characterized by a relatively more holistic stance in comparison with other disciplines, sociology can draw a more panoramic view of tourism by locating it in the wider context of society or the world-system. In so doing, some "sociological imagination" (Mills 1959) is required; imagination allows insight that goes beyond the sight of the empirical eye.

Theorizing is of two types. The first is grounded on *statistical* generalization and inductive logic. The second is based on *analytical* generalization and deductive logic. While it is acknowledged that the two are interrelated, in this book the second type is emphasized from a sociological perspective. As Marx claimed, one cannot use a scalpel or a chemical reagent to study commodities; rather, one must employ abstraction in order to analyze the social relations underlying commodities. The same methodological principle is applicable to tourism. The latter is a social fact, embodying the social relations and cultural norms that underpin it. To discover these, one needs to appeal to theoretical imagination, that is, mental abstraction, in order to analyze these underlying social conditions and relations that are responsible for the emergence of tourism. This theorizing from a sociological approach is not a lone activity. Fortunately, it is possible to consult the literature and mobilize various theoretical and empirical sources in order to reinforce the arguments contained in this volume.

As to the structure of the book, it is organized in terms of its central theme: the structural ambivalence of modernity. It consists of three parts. Part One, featuring one chapter only, is about the differentiation of modernity into Logos-modernity and Eros-modernity, and the ambivalence of modernity is concretized as the ambivalent relationship between the two. Tourism is contextualized in these terms.

Part Two is about the sociological motivations of tourism. The sociogenesis of tourism is explained in terms of the metaphorical "push of modernity", namely the "negative" side of the ambivalence of modernity, especially Logos-modernity. The dark side is concretely exhibited in the institutional, environmental, temporal, and social–spatial dimensions of modernity. The main types of tourism—the tourism of authenticity, nature tourism, holiday-making and international tourism (they overlap with one another)—are all related to these different elements and are examples of the cultural reaction and resistance to the dark side of modernity. This part consists of four chapters, which discuss the relationships between modernity and the four types of tourism mentioned above.

Part Three deals with the social and cultural construction of tourist consumption. Tourism is examined in terms of the metaphorical "pull of modernity", namely the seductive side of modernity, especially touristic consumerism relating to Eros-modernity. This seduction is exhibited not only in the material conditions (such as advances in the technology of transportation and communication, improved living standards and quality of life) and organizational conditions (such as related business organizations or the industry itself) of modernity, but also in its cultural conditions, namely the endless creation of images, signs, discourses, and symbols that constitute the components of consumer culture. While the former are widely acknowledged and well documented, the latter are relatively neglected in the literature. Thus, attention will focus on the latter. In relation to tourism, the cultural seduction of modernity is evident in the domain of tourist images, discourses, consumer culture and tourism-related symbolic goods. Consumption is treated as a specific example of Eros-modernity. However, the supply of consumer services and products involves the organizational form of Logos-modernity, as is shown in related business organizations and the industry itself, which are agencies of the commoditization of touristic experiences and services. Thus, it is difficult to separate the *cultural* production (cultural seduction in marketing, for instance) and the *commercial* production of tourism. Part Three consists of four chapters. These chapters analyse how the tourism production system and society effectively employ cultural seduction in the construction of images, discourses, consumer culture and status-symbols.

This book does not pretend to exhaust all the issues within the sociology of tourism. Rather, a few have been selected for in-depth treatment, namely sociological motivations, as they relate to the structural "push" of modernity, and consumer culture, as it relates to the cultural "pull" of modernity. There are many more areas calling for further sociological investigation, such as the social and cultural consequences of tourism upon destination communities under the conditions of globalization (Britton 1982; Butler and Hinch 1996; Cleverdon 1979; Husbands 1981; de Kadt 1979; Harrison 1992; Lea 1988; Mathieson and Wall 1982; V. Smith 1977, 1989; UNESCO 1976; Young 1973), a topic which is beyond the scope of this book. Even so, the nature of these consequences and their policy implications will be discussed in the conclusion.

PART 1

CONTEXTUALIZING TOURISM

Chapter 2

Logos-modernity, Eros-modernity, and Tourism

As discussed in the preceding chapter, tourism can be regarded as a motivational response to the ambivalence of the existential condition of modernity. It would, however, be misleading to argue that tourism is something that lies "beside" or "outside" modernity. Quite the contrary, it is essentially a modern phenomenon (Cohen 1972, 1995; MacCannell 1976), one that epitomizes modernity. Thus, the issue in question can be further elaborated as "to which part of modernity is tourism responsive?" Even though modernity is often dreary and alienating, it is certainly unfair to claim that every moment of it is tedious. It supplies people with enjoyment, liberty, and hope, as well as malaise, alienation, and frustration. Indeed modernity is essentially ambivalent. However, this ambivalence is asymmetrical with regard to each of its constitutent parts. Some elements of modernity may contain more "hatred" and less "love", while others exhibit more "love" and less "hatred", despite the realization that every part of modernity can be ambivalent. This leads to a further question, namely, which part of modernity reflects more the "hatred" side of the ambivalence and which part more the "love" side? The hypothesis of this chapter is that modernity is structurally differentiated into Logos-modernity and Eros-modernity. Logos-modernity embodies more of the "hatred" than the "love" elements, while Eros-modernity elicits a more joyful response from modern people. Indeed, tourism itself, like leisure, romance, entertainment, and spectator sport, is an example of Eros-modernity. Nevertheless, Logos- and Eros-modernity are both ambivalent. The relationship between them is dialectical and complementary.

The Discourse of Enlightenment and the Logos Version of Modernity

"Modernity" has been one of the most popular terms in use throughout the 1980s and 1990s. Yet it is also a controversial concept, reflected in diverse and competing definitions. While there is no need to go over this in detail here, a general trend can be pointed out. Most versions of modernity seem to mirror the discourse of the Enlightenment, one which supplies a standard version of modernity. Either one can declare the end of modernity and the rise of postmodernity (Lyotard 1984), or one can insist that modernity is an ongoing and incomplete project (Habermas 1985). However, it can hardly be denied that the thinkers of the Enlightenment provided a typical or standard version of the discourse and the terms that can now be used to portray what modernity is. It is, at least in one sense, the *project* of the Enlightenment (Bauman 1987); it is this project in practice.

At the core of Enlightenment thinking are the principles of *reason* and *rationality*. For Enlightenment thinkers, to be modern is to be rational. There is no act or thought that can avoid trial in the court of reason and rationality. Everything must be assessed in terms of these criteria, and everything must be designed according to their principles. Irrational factors such as superstition, stubbornness, ignorance, spontaneity, and so on must be repressed or eradicated. The discourse of the Enlightenment holds to a tradition of Logos that can be traced back to ancient Greek philosophy, particularly to Aristotle. According to this tradition, the essence of humankind is Logos (reason and rationality), and the ideal person is one whose soul is governed and informed by Logos. Thus the project of modernity formulated by the Enlightenment thinkers is a Logos version of modernity; in short, *Logos-modernity*. Modernization, therefore, is in essence rationalization in which Logos has established its hegemony.

The Greek word *Logos*, has a wide range of meanings. According to *The Oxford Companion to Philosophy*, Logos primarily signifies "in the context of philosophical discussion the rational, intelligible principle, structure, or order which pervades something, or the source of that order, or giving an account of that order" (Honderich 1995:511). In his *Nicomachean Ethics*, Aristotle (1925) uses Logos to denote reason, a part of the soul that guides and governs the other part of the soul (the emotions). Hence, in Western thought, Logos acts as a metaphor, referring to reason and rationality. Accordingly, in speaking of Logos-modernity, modern institutions which embody the principles of reason and rationality and exhibit a social order which has its source in reason and rationality, are referred to in Logocentric terms.

The Logos version of modernity is, so to speak, an inheritance of Enlightenment thinking. Since then it has been taken for granted, naturalized, and accepted as the *only* version of modernity. For instance, the classic founders of modern sociology—Weber, Durkheim, and Marx— have each, in various ways, described and examined the processes of modernization in which reason and rationality have established their own hegemony and embodied themselves in the main institutions of modernity, including the bureaucratic state (Weber), industrialism (Durkheim), and capitalism (Marx) (Giddens 1990). For them, what characterizes modernization is rationalization, a process in which reason and rationality overcome irrational factors. Such a Logos version of modernity has also been accepted by contemporary sociologists such as Giddens (1990) and Habermas (1984, 1987). Berger et al characterize rationalizing modernity (i.e., Logos-modernity) in the following terms:

> The rationality that is intrinsic to modern technology imposes itself upon both the activity and the consciousness of the individual as control, limitation and, by the same token, frustration. Irrational impulses of all sorts are progressively subjected to controls.... The result is considerable psychological tension. The individual is forced to "manage" his emotional life, transferring to it the engineering ethos of modern technology (1973:163).

In more general terms, Logos versions of modernity tell the story of how *reason and rationality have gained victory and ascendancy over irrational or non-rational factors, and how irrational or non-rational factors have been successfully subdued, repressed, controlled, or constrained* through the powers of both rational agency (such as modern surveillance and management) and rational mechanisms (such as formal organizations).

Of course, not all irrational and non-rational factors (e.g., emotional factors) are necessarily the target of attack and control by the agents and mechanisms of rationality. Some emotional factors, such as the ideal-typical Protestant's emotional engagement in work, can coexist with rational purposes (Weber 1970). Nevertheless, within the domain of the mainstream institutions of modernity, such as formal organizations, there is little place for irrational and non-rational factors (instincts, impulses, spontaneous drives, passions, and so on). For example Taylorism, a doctrine that has shaped modern managerial culture, declares war on irrational and spontaneous factors in the factory: jobs must be rationally designed, work must be scientifically managed and laborers must be constrained by strict discipline. In other words, irrational and spontaneous drives and emotions must be driven out of the workplace by scientific management.

The Logos version of modernity has celebrated the marvellous material, intellectual, and organizational miracles of Logos found in civilization. However, it is interesting to note that adherents of the Logos version of modernity have also assumed a critical orientation towards Logos-modernity. For example, while Marx, Weber, and Durkheim celebrated the progress of modernization, they also delivered verbal attacks on the limitations of Logos-modernity. Thus, Marx criticized the *alienating* conditions of capitalism, Weber complained of the "*Iron Cage*" of bureaucratic state, and Durkheim revealed the prevalence of *anomie* in industrial society. The tragic consequences of two world wars in this century led to much more serious critiques of Logos-modernity (The Frankfurt School or the School of Critical Theory being perhaps the most typical). These critiques can be summarized as follows:

First, Logos-modernity gradually eliminates the play element from its own space. It is increasingly at odds with the requirements of *Homo Ludens* (Huizinga 1949). Work and life within the sphere of Logos-modernity are, to a certain extent, alienating. Second, characterized by what Taylor (1991) calls the "primacy of instrumental reason", Logos-modernity pursues a given end by the most efficient means available, regardless of more all-encompassing, substantial, and comprehensive goals. Thus, rationality of the *part* often ends up as irrationality of the *whole* (Bauman 1987, 1990b:193–4). For example, the "downside of industrialism became fearfully apparent in the degradation of the environment, the depletion of unrenewable resources, and the deterioration of the ozone layer" (Lyon 1994:6).

Third, Logos-modernity may produce some dehumanizing effects that are opposite to the original goals of reason. In Bernstein's words, "the reason which was to be the means to satisfying human ends becomes its own end, and thereby turns against the true aims of Enlightenment: freedom and happiness" (1991:4). That is to say, Logos tends to achieve its instrumental ends at the expense of the desires of Eros and organic life. Logos is likely to belittle many of the "autotelic" aims and activities which are pursued for their own sake, but which cannot be measured by sheer utilitarian or instrumental criteria (though some of them, admittedly, can coincide with utilitarian purposes) (Csikszentmihalyi 1975). Thus, Fromm's description of the modern economic system also applies to Logos-modernity: this system is "no longer determined by the question: *What is good for Man?* but by the question: *What is good for the growth of the system?*" (1976:16). On the one hand, Logos, in its absolute state (the exclusion of Eros) and taken to its extreme, may produce dehumanizing effects. For example, according to the English social philosopher John Stuart Mill, his highly rationalized training in logic and economics led him to a nervous breakdown. Only by reading Wordsworth's poetry was he able to restore his sanity (see Griswold 1994:5). On the other hand, Logos, as characterized by the "primacy of instrumental reason", may turn against itself. As Lyon puts it: "Reason brought as many nightmares as sweet dreams and the irrationalisms of drugs or new religions promised better" (1994:6), not to mention the tragedy of the holocaust, which in a way was also linked to instrumental rationality and modernity (Bauman 1989).

Indeed, the proponents of the Logos version of modernity (the discursive account of Logos-modernity) do reveal, correctly, the disappointing and dissatisfying aspects of Logos-modernity and therefore provide a normative foundation for future transcendence of its one-sidedness. However, they fail to recognize the actual, practical, or folk resistance to the unidimensionality of Logos-modernity launched by Eros during the course of modernization. The Logos version of modernity has mistaken the "ideal type" of Logos-modernity for the *status quo*. True, Logos dominates modern Western societies. Yet it should be noted that Logos-modernity is first of all an ideal type. In Wagner's words, modernity is "the modern imaginary" propagated by "elite intellectuals" (1994:24). Its existence in the real world is a question of degree, rather than an absolutely full and pure state. " 'Modernity', so to speak, had very few citizens by 1800, not many by 1900, and still today it is hardly the right word to characterize many current practices" (1994:24). In other words, although society can be *to a greater or lesser extent* characterized by Logos, no society in the West has ever been totally governed by Logos in its absolute form. Rather, there has always been some room in Western societies for Logos to concede ground to Eros.

Put another way, the Logos version of modernity is only one of the perspectives used to portray modern society. There is an alternative view from which another side of modern society, a facet that is ignored by most Logos versions of modernity, can be described. This other side is the side of Eros, of the *resistance* coming from Eros (irrational and non-rational desires and demands), and of the compromise made by Logos in relation to Eros. While believers in the Logos version of modernity correctly describe how Logos (reason and rationality) subdued irrational and non-rational factors during the course of modernization, they remain silent on the issue of how irrational and non-rational desires and requirements have been gratified or channelled to approved zones, to be satisfied within the context of modern society. In other words, the Logos version of modernity fails to recognize the Eros dimension of modernity, fails to pay sufficient attention to the carnivalesque, play, romantic, or Dyonisus features of modernity, and consequently fails to attribute enough significance to the studies of entertainment, play, leisure, and tourism. Why did the triumph of industrialization in the eighteenth century almost simultaneously provoke the romantic movement? Why did the spirit of consumerism and hedonism coexist with the Protestant ethic in modern Western societies (Campbell 1987)? Why did modernity witness the enthusiasm for pleasure seeking in the form of institutional leisure (for instance spectator sports, film watching, recreational travel) in addition to disciplined work? Do these factors indicate that there is another dimension to modernity besides the Logos version? To put it another way, is Logos-modernity really the only version of modernity or is there another branch? To answer these questions, inspiration can be drawn from Lefebvre's (1991) discussion of the Logos–Eros polarization.

Lefebvre on "the Great Logos–Eros Dialectic"

In *The Production of Space* (1991), Lefebvre likens the struggle between "abstract space" and "differential space" to the struggle between Logos and Anti-Logos (Eros). As he observes:

> The Logos makes inventories, classifies, arranges: it cultivates knowledge and presses it into the service of power.... On the side of the Logos is rationality, constantly being refined and constantly asserting itself in the shape of organizational forms, structural aspects of industry, systems and efforts to systematize everything, and so forth. On this side of things are ranged the forces that aspire to dominate and control space: business and the state, institutions, the family, the "establishment", the established order, corporate and constituted bodies of all kinds (1991:391–392).

It is quite obvious that the above quotation implies a notion of Logos-modernity. For Lefebvre, the space of modernity is dominated by Logos, which not only ensures access to objective knowledge to create order in the world, but also embodies itself in a number of institutions (business, state, family, and so on). In Lefebvre's terms, Logos-modernity is characterized by "quantitative", "abstract", or "homogenized" space in which order is produced in terms of instrumental reason (Logos). However, thanks to the forces of Anti-Logos or Eros,

> a theatricalized or dramatized space is liable to arise. Space is liable to be eroticized and restored to ambiguity, to the common birthplace of needs and desires, by means of music, by means of differential systems and valorizations which overwhelm the strict localization of needs and desires in spaces specialized either physiologically (sexuality) or socially (places set aside, supposedly, for pleasure) (1991:391).

Thus, at the opposite end of the continuum to Logos, Eros tends to create a "differential space" in opposition to "abstract" or "homogenized space". This differential space is associated with "Grand Desire" (Nietzsche) or Eros. It is a space of "appropriation" rather than "domination" (1991:392). It is qualitative rather than quantitative.

According to Lefebvre, the struggle between Logos and Eros constitutes the great dialectic. This dialectic explains a substantial part of our civilization; for example the contradiction between technology on the one hand and poetry and music on the other:

> Implicit in the great Logos–Eros dialectic, as well as in the conflict between "domination" and "appropriation", is a contradiction between technology

and technicity on the one hand, and poetry and music on the other. A dialectical contradiction, as it is surely needless to recall, presupposes unity as well as confrontation. There is thus no such thing as technology or technicity in a pure or absolute state, bearing no trace whatsoever of appropriation. The fact remains, though, that technology and technicity tend to acquire a distinct autonomy, and to reinforce domination far more than they do appropriation, the quantitative far more than they do the qualitative. Similarly, although all music or poetry or drama has a technical—even a technological—aspect, this tends to be incorporated, by means of appropriation, into the qualitative realm (1991: 392).

Thus, according to the "great Logos–Eros dialectic" there should be two different spaces in modernity. On the one hand, there is the space of Logos-modernity, characterized by "abstract space", in which the quantitative rather than the qualitative, the utilitarian rather than the poetic, the instrumental rather than the substantial, domination rather than appropriation are dominant. On the other hand, there is a "differential space". This is of particular significance because it is the space of Anti-Logos or Eros. If the space of Logos is the key source of order and control, then that of Eros is the origin of life satisfaction. Without it the world would be to a large extent meaningless and life would be extremely dull. In reality, Eros always confronts and resists Logos because Eros is a primitive and dynamic part of human existence. Thus, modernity also includes a "poetic space", a "space of desires", a "space of pleasure", a "space of play"— in short, a "space of Eros". Although such a space remains marginal in modernity, it does represent an alternative (Shields 1991), that is to say, it is a space of *Eros*-modernity.

Eros-version of Modernity

In Greek mythology, the god Eros is in charge of matters of love, but Freud (1963) and Marcuse (1955) use the term Eros to denote "the life instinct". For Marcuse, Eros has two meanings. First, it refers to the organic, instinctual dimension of life, which is the source of pleasure, something more than Freudian sexuality. "Eros, as life instinct, denotes a larger biological instinct rather than a larger scope of sexuality" (1955:147). Second, Eros signifies an ideal state in which the gratification of instinctual pleasure is no longer at odds with reason and rationality. In other words, repressive reason gives way to a new rationality which encourages, rather than restrains, pleasure (1955:158).

Like Freud, Marcuse attaches an ontological meaning to Eros (1955:95). If mainstream Western thought dating from Plato and Aristotle treats Logos as the essence of Being, then Marcuse, following Freud, regards the essence of Being as Eros: "Being is essentially the striving for pleasure" (1955:95). Eros is the builder of culture. The history of civilization is nothing more than an account of the struggle between Eros and Logos (reason), or a tradition of Logos imposing restraints and repression upon Eros. Marcuse (1955:29) represents the history of civilization as a series of transitions: from immediate satisfaction to delayed satisfaction, pleasure to restraint of pleasure, joy (play) to toil (work), receptiveness to productiveness, absence of repression to security.

Civilization refers to the progress of humanity through which instinctual drives and impulses are subdued by Logos. But in the name of reason, civilization also imposes a "surplus-repression" on Eros in the interests of domination. Civilization is thus highly ambivalent. It has made great achievements at the expense of Eros. In *Eros and civilisation,* Marcuse explores the possibility of there being an ideal state in which the instinctual or Eros gratification can converge with (non-repressive) reason, rationality or Logos, in which the surplus repression of Eros is no longer necessary. For Marcuse, the key to this ideal state is constituted by non-repressive conditions:

> Under non-repressive conditions, sexuality tends to "grow into" Eros—that is to say, toward self-sublimation in lasting and expanding relation (including work relations) which serve to intensify and enlarge instinctual gratification (1955:157).

In other words, under the conditions of non-repressive sublimation, sexuality is transformed into Eros (1955:ch.10). Eros thus entails a "non-repressive culture" in which instincts and reason form "a new relation":

> The civilized morality is reversed by harmonizing instinctual freedom and order: liberated from the tyranny of repressive reason, the instincts tend toward free and lasting existential relations—they generate a *new* reality principle (1955: 142).

Therefore, under the ideal conditions of a non-repressive culture (for example the "aesthetic state"), "Eros redefines reason in his own terms" (1955:158). Eros renegotiates and alters the relation between instincts (pleasure principle) and reason (reality principle) and "repressive reason gives way to a new *rationality of gratification* in which reason and happiness converge" (1955:158). Whereas Logos is the reality principle and libido is the pleasure principle (unlimited gratification), Eros is "the principle of Being" (1955:95).

As mentioned above, Marcuse treats Eros as an ideal state of Being under the conditions of a non-repressive culture, in which the gratification of instinctual pleasure is consistent with Logos, and hence allowed, approved or encouraged. This ideal has been criticized by many commentators and dismissed as utopian. However, although Marcuse's Eros-gratification takes on the form of utopia, it does have some grounding in reality. His discussion of Eros gives a hint that there should be an Eros-modernity in which the requirements of reason/rationality and the gratification of Eros converge, "in which reason and happiness converge" (1955:158). As Lefebvre (1991) argues, although Logos dominates the spaces of modernity, there is some evidence to demonstrate that modernity has also supplied Eros with a certain space. In reality, modernity has not only witnessed a process in which rationality has subdued and restrained irrational and non-rational factors (Logos-modernity); it has also involved a process in which certain irrational and non-rational factors (Eros) have been *licensed and channelled to approved, safe, structurally separated zones, to be released and celebrated*, rather than repressed and constrained (Shields 1991). In such "legitimate" zones—though marginal and liminal—irrational and non-rational factors (such as instincts, impulses, desires, drives of play and pleasure, feelings, emotions, imaginations, and so on) are no longer required to be subdued, constrained and controlled, but released, satisfied, or consumed. For example, modernity has seen the emergence of the "privatizing of passion", which offers additional scope for a free emotional life (Giddens 1991); of the institutionalization of "spectator sports", which sets up a safe and harmless channel for the release of violent impulses and emotions that are considered to be dangerous in mainstream institutional areas of civilized societies (Elias 1978; Elias and Dunning 1986); of "free areas" or "escape routes" where the rules of rational self-constraint are relaxed and activities that may not be consistent with the requirements of rationality are permitted (hobbies, games, gambling, sex, holidays, drugs, therapy, and so on; see Cohen and Taylor 1992); of nudity on the beach (so called "dirty places"), which is culturally prohibited by the sober and rational centre of society (Shields 1991). That is to say, in addition to the Logos version of modernity there is an alternative version. While Logos-modernity celebrates reason and rationality, this alternative modernity celebrates biological instincts, impulses, desires, spontaneity, feelings, passions, the imagination—in short, Eros. This alternative modernity can be thus called *Eros-modernity*.

In a strict sense, Eros-modernity is not itself a homogeneous system, despite its goal to serve Eros. For the sake of analytical simplicity, Eros can be further subdivided into *poetic Eros* (cultivated emotional and imaginative pleasure, romanticism, or sublimation of instinctual pleasure in a Freudian sense) and *carnivalesque Eros* (sensual pleasure, sensation seeking, or the Dionysian spirit in a Nietzschean sense). Correspondingly Eros-modernity can be subdivided into Eros-modernity 1 and Eros-modernity 2 (following Rojek's 1995 division of modernity 1 and modernity 2), as follows: Poetic Eros— Eros-modernity 1; carnivalesque Eros—Eros-modernity 2. Associated with poetic Eros, Eros-modernity 1 is mainly about sublimated, cultivated, or "civilized" forms of Eros gratification in a Freudian sense. It seeks Eros gratification in such a way that it can peacefully coexist with Logos (reason and rational order), hence reinforcing the order that Logos requires. Eros thus satisfies itself in a detour.

Eros-modernity 1 channels the energies of Eros into sublimation, into spiritual transcendence, and artistic creative activities: music, poetry, painting, and so on. In other words Eros-modernity 1 mainly consists of cultivated, refined, and sublimated cultures, namely "high cultures" or elitist cultures, including the arts, literature, landscape tourism, and other "pure" cultural or romantic activities. Under modernity, such "high cultures" are the most legitimate zones in which the demands of Eros are satisfied. Although Eros-modernity 1 may still have a problematic relationship with Logos-modernity, it coordinates well with Logos-modernity in maintaining social order. Its major aficionado is the middle class.

Linked to carnivalesque Eros, Eros-modernity 2, in contrast, is mainly about the direct, less-sublimated, less cultivated, even crude and "dirty" forms of gratification of Eros, particularly the sensual aspects of Eros (instinctual pleasure). Carnivals, "dirty" jokes, gambling, erotic and violent literature and films, "sweet talk" on telephone lines, sexual freedom, sex tourism, nudity on sunny beaches, hippies, spectacular "cruel" sports such as boxing, excessive drinking, entertainment, and so on are all the examples of Eros-modernity 2. In these areas Logos recedes and Eros surfaces as the main preoccupation. Rather than seeking gratification as sublimation, carnivalesque Eros demands its own satisfaction in a direct, crude, primitive, or less cultivated way. Pleasure is the only goal. People seek pleasure for pleasure's sake. Nothing else matters. Even TV programmes dedicate a lot of time to sheer pleasure: recent examples in Britain include "Blind Date", "Beadle's About", "You've been framed" and "Seeking Pleasure".

Unlike Eros-modernity 1, Eros-modernity 2 does not sit well with Logos-modernity. Rather, it relaxes the rules of Logos, weakens its rational order, and causes problems for it. However dangerous it may be, Eros-modernity 2 has gained its own approved zones or space, including "low culture", "low leisure", entertainment, and various other forms of pleasure seeking. But due to its potential dangers, Eros-modernity 2 is restricted to structurally separated, marginal, and liminal zones (Shields 1991). In contrast to Logos-modernity, Eros-modernity 2 is on the periphery of modernity. It is frequently signposted by a series of markers.

Eros-modernity 1 can be illustrated more clearly by borrowing some of Shanks' (1992:132) paradigmatic pairs of oppositions, as follows (some inappropriate pairs are omitted):

- Objectivity–Subjectivity
- Abstract–Concrete
- Rationality–Emotion
- Truth–Beauty
- Detached–Involved

These contrasts are helpful in illustrating Eros-modernity 1 in relation to Logos-modernity. To risk oversimplification, it can be said that the left-hand elements are characteristic of Logos-modernity and the elements on the right of the pairings belong to Eros-modernity 1. The former describe the orientation that the agents of Logos-modernity must adopt. To ensure a rational order in relation to nature, society, and the self, the Logos subject (rational humanity) assumes an *objective, rational, and detached* orientation towards the world. Only by overcoming irrational, emotional, or Eros elements are persons able to reach the truth and gain objective (abstract) knowledge. These are the means through which Logos subjects acquire the strength to control the world (Elias 1956).

In contrast, the agents of Eros-modernity 1 assume a different stance toward the world, that is, a viewpoint encapsulated by the right-hand elements: a *subjective, emotional, and involved* orientation. While the agents of Logos-modernity engage in the search for truth, the agents of Eros-modernity are absorbed by the beauty of the world, by its aesthetic, poetic, and affective features, by the concrete forms of reality rather than the abstract laws (Logos) behind reality.

Historically, whereas industrialists are Logos subjects (the agents of Logos-modernity), romantics are typically the agents of Eros-modernity 1. Whereas Enlightenment thinkers celebrate Logos-modernity, romantics laud Eros-modernity 1 (aesthetic, poetic, emotional, or romantic life, i.e., Eros life).

Shields' (1991:260) pairs of oppositions are useful in illustrating Eros-modernity 2 in relation to Logos-modernity:

- Rational–Ludic
- Civilized–Nature
- Centre–Periphery
- Social order–Carnivalesque
- Mundane–Liminal

Again, Logos-modernity is characterized by the elements listed on the left and Eros-modernity 2 by those on the right. Logos-modernity represents the *social order* of modernity since it is a system in which irrational and non-rational factors (instincts, impulses, emotions, and so on) have been subdued, repressed, and controlled by a rational and civilized agency. Logos-modernity is the centre of modern society, both because Logos (reason and rationality) is the central value of modernity and because it constitutes the mainstream institutions of modernity. It is a mundane existence.

Eros-modernity 2, by contrast, is a *carnivalesque* "enclave" in which Eros can gratify itself in a direct, non-distorted, and even crude way. The Eros subject (pleasure seeker) is allowed to return to the *natural* state of humanity (e.g., playing on a sunny beach while on holiday), but only within liminal zones (Lett 1983). Eros-modernity 2 is at odds with the order and rules of reason and rationality (Logos). It is, thus, structurally demarcated and separated from Logos-modernity, and defined as a liminal, although reserved, domain for Eros. Eros-modernity 2 only stands in the margin space of modernity.

From the Economics of Eros to the Politics of Eros

According to Freud (1963), the history of civilization is a story about how the desire of Eros has gradually been moderated, transformed, sublimated, constrained, or had its satisfaction postponed, or about how the "pleasure principle" has been transformed into the "reality principle" (Logos). For Freud, "what decides the purposes of life is simply the programme of the pleasure principle. This principle dominates the operation of the mental apparatus from the start" (1963:13). However, under the pressure of punishment and suffering which come from the external world, "men are accustomed to moderate their claims to happiness—just as the pleasure principle itself, indeed, under the influence of the external world, changed into the more modest reality principle" (1963:14). The reality principle thus constitutes the economics of pleasure (Eros). It subdues and constrains the unlimited demands of the pleasure principle (libido) and instead seeks an economic approach towards libido gratification in order to avoid pain. Therefore, "Happiness, in the reduced sense in which we recognise it as possible, is a problem of the economics of the individual's libido" (1963:20). A part of the instinctual drive for pleasure is then abandoned to the avoidance of suffering and punishment. "Civilized man has exchanged a portion of his possibilities of happiness for a portion of security" (1963:52). The economies of pleasure lead to the transformation of Western subjectivity in the sense that, during the course of the civilizing process, Westerners have established internal mechanisms of self-constraint in their psychic structures. This situation, to certain extent, corresponds with the process of state formation, in which the means of violence have been gradually monopolized by the centralized state (Elias 1978, 1982).

Whereas the reality principle refers to the *economics* of pleasure (Eros) in terms of an *individualistic* perspective, the same principle indicates the *politics* of pleasure (Eros) in terms of an *holistic* perspective. If the economics of Eros show people how they can gratify their Eros demands without causing suffering and punishment, then the politics of Eros determine how, when, and where a pleasure-seeking activity is acceptable to a society, particularly to the state.

Every society has its own mode of Eros. From carnivals in medieval society to contemporary narcissistic and hedonistic culture in "the permissive society", Eros gratifies itself in a concrete, culturally, and socially approved and accepted form, and a socially and historically specific form. Civilization is not about whether or not Eros should be satisfied but rather about how and to what extent Eros can be gratified.

Seen this way, modernity not only tells a story of how Logos governs society, but also provides an account of how Eros is channelled into approved or tolerated zones where it can be satisfied. To risk oversimplification, if there is no clear structural separation between Logos and Eros in traditional society (just as there is no institutional separation between work and leisure), then the structural and institutional separation of Eros from Logos is as much a characteristic of modernity as is the institutional differentiation between work and leisure. That is to say, institutional modernity consists of not only Logos-modernity but also Eros-modernity. Whereas industrial, capitalist, commercial, and bureaucratic institutions are the sites where Logos-modernity is located, the institutions of leisure and culture are the locations where Eros-modernity resides, though leisure and culture are not exclusively Eros-oriented. As Elias and Dunning (1986) observe, leisure itself is a spectrum, consisting of many layers of activities, some of which could be Logos-oriented.

Ideal-typically speaking, in a society in which there is no structural and institutional separation between Logos and Eros, Eros either lets itself go and permeates Logos, or else faces violence and coercion from divine powers or traditional taboos. In contrast, under modernity the structural and institutional differentiation between Logos and Eros helps to gratify Eros in a novel way. That is to say, within certain zones Eros can satisfy itself without necessarily threatening the order and institutions of Logos, because Eros is kept out of the mainstream institutions of Logos and channelled into structurally separated areas. Although there are tensions between Logos-modernity and Eros-modernity, both Logos and Eros seem to coexist peacefully—so-called organic integration in a Durkheimian sense.

The organic integration of Logos and Eros in modernity is exhibited in the fact that Eros-modernity helps reinforce the order of Logos-modernity. One of the consequences of modernization is secularization (the weakening of sacred roles and religious integration), a situation conducive to anomie under modernity. It seems to be the case that desires (consumer culture) and Eros (pleasure, leisure, narcissism, hedonism) have replaced religion in support of the order of Logos-modernity. Whereas religion calls for the restraint of desire, Eros-modernity stirs up desire and promises a wider variety and more scope for Eros gratification. By offering greater opportunity for gratification and more sensual pleasures to the masses, Eros-modernity manages to persuade people to forget the alienation caused by Logos-modernity, and hence to work together with Logos-modernity to reproduce the social order needed by Logos. As Horkheimer and Adorno put it, "Amusement under late capitalism is the prolongation of work. It is sought as an escape from the mechanized work process, and to recruit strength in order to be able to cope with it again" (1973:137).

Another fact that signifies the organic integration of Logos and Eros in modernity is that both Logos-modernity and Eros-modernity enhance each other through mutual support. First, there is the incorporation of Eros elements into Logos-modernity. Here Logos-modernity often tries to make use of Eros as an integrative element in its own system. For example the Eros-element of irrational desire, romantic feelings, erotic passions, and so on are increasingly employed in advertisements which aim to arouse desires on the part of consumers, and hence to further market interest. The continuous expansion of production is matched by ever increasing desires and dreams, which contain romantic or Eros elements (Campbell 1987). Recent studies of consumer culture also confirm the economic implications of the Eros element. Second, Logos plays a significant role in supplying Eros objects. In the context of modern society, one of the characteristics of Eros-modernity is the *privatising* of Eros. This state of affairs is perhaps one of the reasons why contemporary Western society is called the "permissive society". Eros demands are, in many cases, personal and private affairs, hobbies, or the quest for freedom. Another characteristic of Eros-modernity is that the supply of Eros objects, or the gratification of Eros demands, is often realized with the help of Logos, i.e., the *commodification* of Eros. Thanks to the help of Logos, Eros objects are supplied in an efficient way, on a massive scale, and in a standard form.

Contextualizing Tourism

Generally speaking, the history of civilization is a history of the "great Logos–Eros dialectic". Whereas Eros is a primitive source of pleasure, happiness, and life satisfaction, Logos is a source of order. Although Logos and Eros are opposite to each other, they are also dependent on one another. On the one hand, without being limited and guided by Logos, Eros can still bring about punishment and suffering. On the other hand, Logos is a *means* to happiness. However, if Logos is taken to an extreme and seen as an ultimate end rather than a means, then it will turn towards the opposite of its primary goal—dehumanization rather than happiness.

Logos–Eros polarization is a universal phenomenon, specifically demonstrated in the dichotomies between work and leisure, business and play, rationality and emotion, truth and beauty, labour disciplines and consumer culture, system and culture, social order and the carnival-esque, and so on. What is specific to modernity is that both Logos and Eros have been institutionalized as two separate and demarcated domains—Logos-modernity and Eros-modernity. This institutionalized differentiation between Logos and Eros is followed by a shift from the economics of Eros to the politics of Eros. If the economics of Eros is an individual's strategy of gratifying Eros without being punished, then the politics of Eros is a strategy of modernity as a whole, which channels and licenses the demands of Eros (irrational and non-rational factors) to the socio-politically approved and socio-culturally acceptable zone of gratifi-cation. Thus, the Eros gratification in Eros-modernity no longer threa-tens the order of Logos-modernity, where irrational and non-rational factors (Eros) must be subdued, constrained, or driven out. Rather, Eros gratification turns to support and reinforce the order of Logos-modernity through pleasure seeking, a situation which helps people for-get the malaise brought about by Logos-modernity.

Modern tourism, as institutionalized leisure travel, is one of examples of Eros-modernity that allows people to gratify their Eros impulses and desires without being punished by the agents of Logos. From the parti-cipant's perspective, tourism is essentially a non-productive pursuit, an activity of sheer consumption. However, it is a series of experiences approved by the tourist-sending society. Moreover, a person's Eros impulses and desires can be gratified or released in tourist activity. For example tourists can give rein to spontaneity while on holiday: eat when they are hungry, sleep when they are tired, make love *ad libitum*, rather than doing these things in accordance with the dictates of sociability and reason (Gottlieb 1982). In tourism settings tourists are removed from the space of factories, offices, or other places of work, where Logos con-strains, supervises, and controls the spontaneity and irrational impulses of Eros. Thus, in tourist space, tourists to a large extent act in terms of the principle of Eros gratification rather than the principle of self-constraint by Logos. The gratification of Eros in and through tourism, then, releases the tensions caused by the self-constraints imposed by Logos on Eros. In this way, tourism helps reinforce the order of the home society that Logos underpins. In other words, the temporary touristic "deviance" or "escape" helps to restore the "normality" of society at home (Urry 1990a). Tourism, so to speak, as a kind of Eros-modernity, coordinates with Logos-modernity organically.

Unlike tourism as pleasure travel or a leisure institution, the tourism industry, as an agent for the commodification of touristic experiences, is the embodiment of Logos. While tourists are largely motivated by Eros, the tourism industry is to a large extent informed by Logos. Whereas tourists are able to pursue their Eros gratifications, employees in the industry must restrain their impulses and emotions, or Eros—they must work in terms of the dictates of Logos. Therefore, if tourism as pleasure travel is an example of Eros-modernity, then the tourism industry is the opposite. The latter is rather an instance of Logos-modernity. However, although both of them can converge with one another, tensions frequently arise. Whereas the tourism industry is profit driven and Logos informed, the tourist is romance driven and Eros dominated. Thus, while tourists seek romantic, authentic, and exotic experiences, the industry often provides them with standardized, contrived, and insulated experiences, for it performs its job in terms of the principle of Logos, i.e., "efficiency, calculability, predictability, and control" (Ritzer 1996). As a result, the relationship between the tourist experience (e.g., Eros-modernity) and the tourism industry (e.g., Logos-modernity) is one of ambivalence.

In summary, modern tourism, as both an experience (i.e., Eros-informed institution) and an industry (Logos-informed institution) is a microcosm of the differentiation between Eros-modernity and Logos-modernity that takes place in a wider societal context. However, tourism must also be understood against the larger social context in which Eros-modernity is structurally differentiated from Logos-modernity. In the following two sections, tourism is examined from two points of view. First, tourism is treated as a cultural and motivational response to the structural "push" of the larger Logos-modernity (modern mainstream institutions, technological environments, social rhythms, and social space) (Part Two). Second, tourism is regarded as culturally complying with the "pull" of the tourist industry, a miniature Logos-modernity. The tourism industry constantly resorts to Eros elements in order to offer products and experiences that appeal to tourists who pursue Eros-consumption (Part Three).

PART 2

MODERNITY, TOURISM AND MOTIVATIONS

Part Two of this book is about the relationship between modernity and tourist motivations and experiences. This section concerns the sociogenesis of tourism (in the sense of the "demand side" definition, see Introduction). It is argued here that tourism is not something given, but rather is socially and culturally produced, constructed, and generated. The emergence of tourism is a part of an overall historical process. In premodern times only a small élite could afford to indulge in travel for pleasure—tourism was a privilege for the nobility and the wealthy. It was only under modernity, particularly late modernity, that tourism arose as a need for the majority. Therefore, the emergence of the tourist or mass tourism can only be understood within the context of modernity, particularly late modernity.

The need and demand for tourism can be explained from a number of perspectives. The first focuses on economic and technological explanations. That is to say, a number of the economic and technological factors that help directly to activate tourist motivations and demands are identified—for example an increase in disposable income and personal time, advances in the technologies of transportation and communication, and so on. This approach helps to answer the question "*why* do individuals travel?" in terms of enabling socioeconomic conditions, the so-called economic "pull of modernity". Whereas most tourism textbooks include this perspective, they often fail to state *what* people actually want and do in their pleasure travels.

A second perspective focuses on psychological or social-psychological explanations of tourist motivations and behaviours. They seek to explain not only why individuals travel in terms of universal biological and psychological make-up, factors, or conditions (the psychological approach), but also what they actually want and do in their leisure travels. The social-psychological approach also incorporates certain sociological profile categories such as age group, social class, gender, lifestyle, and so on, into its explanation of tourist motivations and behaviors. These two approaches have overcome the shortcomings of the *external* explanations of the economic/technological approach, which ignores the subjective and psychological factors of tourist need. However, the psychological and social-psychological approaches seem to be ahistorical, failing to take into account the processes through which tourism arises as a collective fact or "total social fact" (Lanfant 1989).

A third perspective focuses on sociological explanations. Although both the economic/technological approach and the (social) psychological approach have identified many direct and concrete influences on tourist motivation and need, the sociological approach goes one step further by identifying a deeper reason for tourist motivation emerging as a "social fact". Sociology relates tourism to the larger context, where it arises as a mass phenomenon. Even so the sociological approach is not homogeneous. It consists of many sub-approaches. One of the latter is that tourism, as a total social and historical phenomenon, should be examined within the context of modernity, or in terms of the historical process of modernization. This is the approach adopted in Part Two.

A number of conventional sociological categories are useful in understanding tourist motivation, such as "social class", "lifestyle", "gender", "age group", "status", "sub-culture", "ideology", "hegemony", and so on. However these categories seem to emphasise the *variety* or the *nature* of tourist motivation rather than its *socio-genesis* as a social fact. Although a demographic approach helps to describe collective tourist demand as a social fact, it cannot offer a structural explanation of the sociogenesis of tourist motivation. While these categories are very useful in the sociological study of tourism, not all of them are referred to in this book. This omission does not mean that they are not important, rather it indicates a lack of space for their full examination. Thus, Part Two focuses on the relationship between modernity and the socio-genesis of tourist motivation and experiences, leaving other sociological treatments of tourism to be studied elsewhere.

The sociological approach does not oppose or exclude other disciplinary approaches such as the economic/technological approach, and the (social) psychological approach. Rather, all of them complement one another. Each analyzes tourism or tourist motivations and needs from different angles. Each is partial and needs to be complemented by other disciplinary approaches. The sociological approach is, however, of particular significance in deepening and widening the understanding of tourism, a phenomenon which has for long suffered the stigma of triviality and superficiality.

Methodologically speaking, as Vukonić notes, "the motivation of tourist movements is a highly complex, stratified and multidimensional problem" (1996:42). Tourism is not generated by just one single motive; many other motives may be involved, although the tourist may not be aware of them. In fact, the motives that a tourist states rarely represent all the motives for travel. Other unconscious motives, as Freud suggests, play a role in influencing a tourist's decision to travel. Tourist motivations are thus stratified into several levels. Some are obvious and can be easily identified; others may be quite opaque to the tourist and may need to be revealed by experts. Some are in the realm of the conscious; others are located in unconsciousness or preconsciousness. Some are individually distinctive; others are socially and culturally commonly determined. Tourist motivations are also influenced variously by different conditions. For low-income groups, economic and price considerations are the most important factors. For high-income groups, by contrast, affordability may be simply taken for granted and cultural conditions may play a greater role in shaping their decisions. In short, tourist motivations result from a combination of various factors. It is hardly possible to identify all of them in this short study.

This book chooses to identify the most general existential conditions of modernity that all members of society encounter, conditions that shape the sociogenesis of tourist motivation in a most general, yet ignored, way. It is not claimed that the total truth underpinning tourist motivation has been found—rather it is acknowledged that only a partial glimpse has been achieved.

Part Two relates tourist motivation to the context of modernity, itself a multi-dimensional complex. Accordingly, tourism is classified into different, sometimes overlapping, concrete forms. For the sake of analytical simplicity, each form of tourism is related to a particular dimension of modernity, though it cannot be denied that any given form of tourism may also have relationships with other dimensions of modernity. Thus, Chapter 3 deals with the relationship between the *institutional* dimension of modernity and the *tourism of authenticity*. Chapter 4 tackles the relationship between the *technological* dimension of modernity and *nature tourism*. Chapter 5 discusses the relationship between the *temporal* dimension of modernity and *holiday-making*. Finally, Chapter 6 examines the relationship between the *spatial* dimension of modernity and *international* tourism.

Chapter 3

Modernity and the Tourism of Authenticity

About two decades ago MacCannell (1973, 1976) introduced the concept of authenticity to the sociological study of tourist motivations and experiences. Since then authenticity has become an important item on the agenda of tourism research, and there has been a parallel growth of literature on this issue (Brown 1996; Bruner 1989, 1994; Cohen 1979a, 1988b; Daniel 1996; Ehrentraut 1993; Harkin 1995; Hughes 1995; Littrell, Anderson and Brown 1993; Macdonald 1997; Moscardo and Pearce 1986; Pearce and Moscardo 1985, 1986; Redfoot 1984; Salamone 1997; Selwyn 1996a; Shenhav-Keller 1993; Silver 1993; Turner and Manning 1988; Wang 1997a;). However, with the concept of authenticity being widely used, its ambiguity and limitations have been increasingly exposed. Critics have questioned its usefulness and validity because many tourist motivations or experiences cannot be explained solely in terms of the conventional concept of authenticity. Tourist activities such as visiting friends and relatives, beach holidays, ocean cruising, nature tourism, trips to Disneyland, and travel for special interests such as shopping, fishing, hunting, sports, and so on have little to do with authenticity in MacCannell's sense of the term (Schudson 1979; Stephen 1990; Urry 1990a). As Urry observes, "the 'search for authenticity' is too simple a foundation for explaining contemporary tourism" (1991a:51).

Of course, authenticity is still relevant to certain types of tourism, such as ethnic, historical or cultural tourism, all of which involve some kind of presentation or representation of the Other or of the past. Yet if the concept of authenticity is of limited applicability, how can it be of central importance in tourism studies? Can researchers continue to employ it while ignoring its associated difficulties? Should they discard it altogether, or should they redefine its meaning in order to justify and enhance its explanatory power? This chapter concentrates on the third of these options, namely rethinking the meaning of "authenticity" in terms of existential philosophers' use of the expression. While the two conventional meanings of authenticity in the literature—objective authenticity and constructive authenticity—are discussed, the third usage—existential authenticity—is suggested as an alternative.

This chapter has two aims. First, three different approaches—objectivism, constructivism, and postmodernism – to the issue of authenticity in tourism are reviewed and analyzed. As a result, three different types of authenticity—objective, constructive (or symbolic), and existential authenticity - are clarified. Second, it is suggested that, under the condition of postmodernity, both objective authenticity and constructive authenticity, as object-related authenticity, can only encompass a limited range of tourist experiences, whereas existential authenticity, as activity-related authenticity, is germane to the understanding of a greater variety of experiences. Existential authenticity is further classified into two dimensions—intra-personal and inter-personal authenticity.

In the four sections which follow, the first reviews and analyzes the literature on authenticity in tourism. On the basis of this overview, the second section suggests that existential authenticity offers an alternative perspective. Existential authenticity is also defined and conceptualized in relation to certain kinds of tourism that cannot be properly explained by the conventional model of the "search for authenticity". The third section discusses the touristic concern for authenticity in the wider context of institutional modernity. Finally, in the fourth section, the concrete forms of existential authenticity—intra-personal and inter-personal authenticity—are discussed.

Different Approaches to Authenticity in Tourism

"Authenticity" is a term that is used in so many different senses and contexts that it has become difficult to define (Golomb 1995:7). According to Trilling, its original usage was in the context of the museum,

> where persons expert in such matters test whether objects of art are what they appear to be or are claimed to be, and therefore worth the price that is asked for them—or, if this has already been paid, worth the admiration they are being given (1972:93).

The term was also borrowed to refer to human existence and "the peculiar nature of our fallen condition, our anxiety over the credibility of existence and of individual existence" (1972:93). For example,

Rousseau used the word "authenticity" to refer to the existential condition of being, and he regarded society as the major cause that destroyed it.

However, it is mainly the museum-linked usage of authenticity that has been extended to tourism. For example, the products of tourism, such as works of art, festivals, rituals, cuisine, dress, and so on, are usually described as "authentic" or "inauthentic" in terms of the criterion of whether they are made or enacted "by local people according to custom or tradition". In this sense, "authenticity connotes traditional culture and origin, a sense of the genuine, the real or the unique" (Sharpley 1994:130). However, the application of the museum-linked usage of the term to tourism oversimplifies the complex nature of authenticity in tourist experiences. First of all, authenticity in tourism can be differentiated into two separate issues—the authenticity of tourist *experiences* (or authentic experiences) and that of *toured objects*. These quite separate aspects are often confused as one. Handler and Saxton note this distinction when they point out that "An authentic experience ... is one in which individuals feel themselves to be in touch both with a 'real' world and with their 'real' selves" (1988:243) . Selwyn (1996a) goes a step further by linking the experience of a "real" world to "authenticity as knowledge", namely "cool" authenticity, and in relating the experience of a "real" self to "authenticity as feeling", namely "hot" authenticity. However, it would be wrong to suggest that the *emotional* experience of the "real" self ("hot authenticity") necessarily entails, coincides with, or results from the *epistemological* experience of a "real" world out there ("cool authenticity"), as if the latter were the sole cause of the former. As will be shown, differentiation of "the authenticity of experiences" from "the authenticity of toured objects" is crucial for introducing "existential authenticity" as an alternative source of authentic experiences in tourism. Certain toured objects, such as nature, are in a strict sense irrelevant to authenticity in MacCannell's terms. However, nature tourism is surely one of the main ways of experiencing the "real" self. That is to say, nature tourism implies an existential authenticity rather than the authenticity of objects.

Second, the complex nature of authenticity in tourism is evident from the fact that it can be further classified into three different types—objective, constructive, and existential authenticity (Table 3.1). *Objective* authenticity is linked to the museum usage of the term. It refers to the authenticity of the *original* that is also the toured object. It follows that the authenticity of tourist experience depends on the toured object being perceived as authentic. In this way of thinking, an *absolute* and *objective* criterion is used to measure authenticity. Thus, even though tourists themselves may think that they have had an authentic experience, it can still be judged as *in*authentic, given that many toured objects are in

Table 3.1. Three Types of Authenticity in Tourism Experiences

Object-related authenticity	Activity-related authenticity
Objective authenticity refers to the authenticity of originals. Correspondingly, authentic experiences in tourism are equated to an *epistemological* experience (i.e., cognition) of the authenticity of originals.	**Existential authenticity** refers to a potential existential state of Being that is to be activated by tourist activities. Correspondingly, authentic experiences in tourism are to achieve this activated existential state of Being within the liminal process of tourism. Existential authenticity has little to do with the authenticity of toured objects
Constructive authenticity refers to the authenticity projected onto toured objects by tourists or tourism producers in terms of their imagery, expectations, preferences, beliefs, powers, etc. There are various versions of authenticity regarding the same objects. Correspondingly, authentic experiences in tourism and the authenticity of toured objects are constitutive of one another. In this sense the authenticity of toured objects is in fact a symbolic authenticity.	

fact false and contrived, or form part of what MacCannell (1973) calls "staged authenticity".

Constructive authenticity is the result of social construction. Authenticity is not seen as an objectively measurable quality. Things appear authentic not because they are so but because they are constructed as such in terms of social viewpoints, beliefs, perspectives, or powers. Authenticity is thus relative, negotiable (Cohen 1988b), contextually determined (Salamone 1997), and even ideological (Silver 1993). It can be the projection of dreams, stereotyped images and expectations onto toured objects (Bruner 1991; Silver 1993). In this sense, what the tourist seeks are signs of authenticity or *symbolic* authenticity (Culler 1981).

Unlike objective and constructive (or symbolic) authenticities, which relate to whether and how toured objects are authentic, *existential* authenticity comprises personal or intersubjective feelings that are activated by the liminal process of tourist behaviors. In such liminal experiences, people feel that they are *themselves* much more authentic and more freely self-expressed than they are in everyday life, not because the toured objects are authentic, but rather because they are engaging in non-every-day activities, free from the constraints of daily life. Thus, analytically speaking, in addition to objective and constructive authenticities, existential authenticity is a distinctive source of authentic experiences in tourism. Unlike object-related authenticity, which is an attribute, or a projected attribute, of objects, existential authenticity is a *potential* existential state of Being which is about to be activated by tourist activities. In this sense, it can also be understood as a variant of what Brown (1996) calls an "authentically good time". Existential authenticity, as activity-related authenticity, is thus logically distinguishable from object-related authenticity (see Table 3.1).

The Approach of Cognitive Objectivism: Authenticity as the Original

In his nostalgic critique of mass tourism in terms of heroic travel in the past, Boorstin (1964) condemns mass tourism as a collection of "pseudo-events", which are brought about by the commoditization of culture and the associated homogenization and standardization of tourist experiences. For Boorstin, under commoditization not only are tourist attractions contrived scenes or pseudo-events, but also the "tourist seldom likes the authentic ... product of the foreign culture; he prefers his own provincial expectations" (1964:106). The tourist is thus gullible; "he is prepared to be ruled by the law of pseudo-events, by which the image, the well-contrived imitation, outshines the *original*" (1964:107; emphasis added). Clearly Boorstin's concept of "pseudo-events" implies a notion of objective authenticity. Authenticity is thus the authenticity of the "original", and tourist experiences are kinds of pseudo-events because they are seldom able to see through the inauthenticity of contrived attractions (for a similar view see Dovey 1985; Fussell 1980).

Whereas Boorstin scorns mass tourism and mass tourists, his critics, such as MacCannell, restore sacredness and quasi-pilgrimage significance to the motivations of tourists. Based on Goffman's (1959) differentiation of the "front region" from the "back region", MacCannell points out that the "concern of moderns for the shallowness of their lives and inauthenticity of their experiences parallels concerns for the sacred in primitive society" (1973:589–590). It is thus justifiable for tourists to "search for authenticity of experience" (1973:589). However, according to MacCannell (1973, 1976), there is increasingly a contradiction between the tourist's demand for authenticity (related to a back region) and the *staged* authenticity in tourist space.

It is always possible that what is taken to be entry into a back region is really entry into a front region that has been totally set up in advance for touristic visitation (1973:597).

However, as Selwyn (1996a:6-7) indicates, MacCannell uses "authenticity" in two different senses: authenticity as feeling and authenticity as knowledge. Indeed, when MacCannell points out that tourism involves "the search for authenticity of experience" or for "authentic experience", his tourists are concerned about the state of their authentic feelings. Yet when he refers to "staged authenticity", his tourists turn to a quest for the authenticity of originals, and consequently become the victims of staged authenticity. Thus, their experiences cannot be considered authentic even if they themselves think they have had authentic experiences. What is implied here is a conception of objective authenticity (a similar view on staged authenticity can also be found in Duncan 1978).

Both Boorstin and MacCannell insist on a museum-linked and cognitive objectivist conception of authenticity, in their references to pseudo-events or staged authenticity. The touristic search for authentic experiences is thus no more than an epistemological experience of toured objects which are found to be authentic. The key point at issue is, however, that authenticity is not an either/or matter, but rather involves a much wider spectrum, rich in ambiguity. That which is judged to be inauthentic or staged authenticity by experts, intellectuals, or élites may be experienced as authentic and real from an "emic" perspective. Indeed, this may be the very way that mass tourists experience authenticity. Thus, a revisionist position occurs in response to the complex and constructive nature of authenticity, that is, constructivism.

The Approach of Constructivism: Authenticity as Construction

From the approach of constructivism, to view authenticity as an original or the attribute of an original is too simplistic to capture the complexity of authenticity. Thus, authenticity in MacCannell's sense has been questioned by many commentators (Bruner 1989:113; Cohen 1988b:378; Handler and Linnekin 1984:286; Lanfant 1989:188; Spooner 1986:220-1; Wood 1993:58). According to Bruner (1994), authenticity has four different meanings. First, it refers to the "historical verisimilitude" of representation, namely authentic reproduction which resembles the original and thus looks credible and convincing. For instance, the 1990s New Salem resembles the 1830s New Salem where Abraham Lincoln lived. Second, authenticity means genuine, historically accurate, and immaculate simulation. Authenticity in both the first and the second senses involves the nature of a copy or reproduction rather than the original. Museum professionals use authenticity primarily in the first sense, but sometimes in the second. Third, authenticity "means originals, as opposed to a copy; but in this sense, no reproduction could be authentic, by definition" (Bruner 1994:400). Finally, in the fourth sense authenticity refers to the power which authorizes, certifies, and legally validates authenticity. For example,

> New Salem is authentic, as it is the authoritative reproduction of New Salem, the one legitimized by the state of Illinois. There is only one officially reconstructed New Salem, the one approved by the state government (1994:400).

Thus, as authenticity involves a range of different meanings, to confine authenticity to an original is oversimplistic. As a response and revision, the disciples of constructivism treat authenticity as social construction.

Constructivism is not a coherent doctrine. It is sometimes used interchangeably with "constructionism". While constructionism shares with constructivism most connotations, the former stresses the *social or intersubjective process* in the construction of knowledge and reality, and is often used in conjunction with "social", i.e., social constructionism (Berger and Luckmann 1971; Gergen 1985; Gergen and Gergen 1991). For the sake of simplicity, in the discussion below constructionism will be treated as a sub-perspective within the general perspective of constructivism. There is no space here to outline the history of constructivism and its variants. However, certain basic characteristics of constructivism can be identified (for a detailed discussion see Schwandt 1994). First, its main ontological assumption is that "there is no unique 'real world' that pre-exists and is independent of human mental activity and human symbolic language" (J. Bruner 1986; quoted in Schwandt 1994:125). Reality is rather better viewed as the result of the many versions of human interpretation and construction. It is thus pluralistic and plastic. Second, constructivists hold to a pluralistic and relativist epistemology and methodology. It is claimed that the validity of knowledge is not to be found in the relationship of correspondence to an independently existing world. On the contrary, "what we take to be objective knowledge and truth is the result of perspective. Knowledge and truth are created, not discovered by mind" (Schwandt 1994:125). For constructivists, multiple and plural meanings of and about the same things can be constructed from different perspectives, and humankind may adopt different constructed meanings depending on the particular contextual situation or its intersubjective setting.

This general constructivist perspective is applied to the issue of authenticity by Bruner (1994), Cohen (1988b), Hobsbawn and Ranger (1983). Bruner (1994:407) clearly labels his treatment of authenticity as a constructivist perspective. Although there are differences among the adherents of constructivism, a few common viewpoints on authenticity in tourism can be identified, as follows.

First, there is no absolute and static original or origin upon which the absolute authenticity of originals relies. "We all enter society in the middle, and culture is always in process" (Bruner 1994:407). Second, as the approach of the "invention of tradition" (Hobsbawn and Ranger 1983) shows, origins and traditions are themselves invented and constructed in terms of living contexts and the needs of the present. Furthermore, the construction of traditions or origins involves power and is hence a social process. As Bruner puts it, "No longer is authenticity a property inherent in an object, forever fixed in time; it is seen as a struggle, a social process, in which competing interests argue for their own interpretation of history"(1994:408).

Third, authenticity or inauthenticity is a result of how persons see things and of their perspectives and interpretations. Thus, the experience of authenticity is pluralistic, relative to each type of tourist, who may have their own way of definition, experience, and interpretation of authenticity (Littrell et al 1993; Pearce and Moscardo 1985, 1986; Redfoot 1984). In this sense, if mass tourists empathically experience toured objects as authentic, then their viewpoints are real in their own right, regardless of whether experts propose an opposite view from an "objective" perspective (Cohen 1988b).

Fourth, with respect to the different cultures or peoples that are to be toured, authenticity is a label attached to toured cultures in terms of the stereotyped images and expectations held by members of a tourist-sending society. Culler (1981) demonstrates this point from a semiotic perspective. For example, real Japaneseness is what has been marked as such; however, what is located in Japan without being marked is in a sense not real Japaneseness and hence is not worth seeing (1981:133). Authenticity is thus a *projection* of tourists' own beliefs, expectations, preferences, stereotyped images, and consciousness onto toured objects, particularly onto toured Others (Adams 1984; Bruner 1991; Duncan 1978; Laxson 1991; Silver 1993). As Bruner puts it, tourists' authentic experiences are not based on any real assessment of natives, such as New Guineans, but rather "a projection from Western consciousness" (1991:243). "Western tourists are not paying thousands of dollars to see children die in Ethiopia; they are paying to see the noble savage, a figment of their imagination"(1991:241).

Fifth, even though something in the beginning may be inauthentic or artificial, it can subsequently become "emergent authenticity" as time goes by. Such is the case with Disneyland or Disney World in the United States (Cohen 1988b:380). The infinite retreat of the "now" will eventually make anything that happens authentic. Authenticity is thus an emerging process. It is also context-bound. In an examination of the two San Angel Inns, the original in Mexico City and its "daughter" at Disney World in Florida, Salamone (1997) claims that both versions of the inn are authentic, each in its own way, and each makes sense within its own context.

In effect, for constructivists, tourists are indeed in search of authenticity; however, what they seek is not objective authenticity (i.e., authenticity as originals) but a *symbolic* authenticity which is the result of social construction. Toured objects or toured others are experienced as authentic not because they are originals or reality but because they are perceived as the signs or symbols of authenticity (Culler 1981). Symbolic authenticity has little to do with reality. It is more often than not a projection of certain stereotypical images held and circulated within tourist-sending societies, particularly within the mass media and the promotional tourism materials of Western societies (Britton 1979; Silver 1993).

The Approaches of Postmodernism: the End of Authenticity?

Postmodernism is not a single, unified, and well-integrated approach. Rather a diversity of postmodern views or approaches exist (for a detailed discussion see Hollinshead 1997). However, as regards authenticity in tourism the approaches of postmodernism seem to be characterized by the deconstruction of authenticity. Whereas modernist researchers such as Boorstin (1964) and MacCannell (1973, 1976) worry about pseudo-events or staged authenticity in tourist space, postmodernist researchers do not consider inauthenticity as problematic at all.

Eco's (1986) writing on "hyperreality" represents a typical postmodernist position on the issue of authenticity in tourism. Indeed, Eco totally deconstructs the concept of authenticity through destructuring the boundaries between the copy and the original, or between sign and reality, boundaries upon which the whole issue of Boorstin's and MacCannell's objective authenticity relies. For Eco, the most typical model of hyperreality is illustrated by the example of Disneyland in the United States, for Disneyland was born out of fantasy and imagination. Thus, it is irrelevant whether it is real or false, since there is no original that can be used as a reference.

Based on Eco's idea of "hyperreality", the French postmodernist writer, Baudrillard (1983) borrows the concept "simulacrum" from Plato to explain different cultural orders in history. According to Baudrillard (1983:83) there are three historical "orders of simulacra" which refer to different relationships between simulacra and "the real". The first order of simulacra emerges in the period from the Renaissance to the beginning of the industrial revolution. The dominant simulacrum of this period is "counterfeit", which indicates the emergence of representation. The second order of simulacra—"production"—appears in the industrial era, which indicates the potential for exact technical reproduction and reproducibility of the same object. The third order of simulacra is simulation, which refers to the contemporary condition. In the postmodern world individuals "live by the mode of referendum precisely because there is no longer any referential" (1983:116). "The contradictory process of true and false, of real and the imaginary, is abolished" (1983:122). The world is a simulation which admits no originals, no origins, no "real" referent but the "metaphysic of the code" (1983:103). Like Eco, Baudrillard also uses Disneyland as a prime example of simulation (1983:23).

In a discussion of the culture of Disney, Fjellman claims that:

> The concepts of real and fake, however, are too blunt to capture the subtleties of Disney simulations. At WDW things are not just real or fake but real real, fake real, real fake, and fake fake (1992:255).

Therefore in Disneyworld there is no absolute boundary between the real and the fake. The real may turn into the fake and vice versa. The "Disney plan is to juxtapose the real and the fake", and the "lines between the real and the fake are systematically blurred" (1992:255).

Implied in the approaches of postmodernism is justification of the contrived, the copy, and the imitation. One of the most interesting responses to this postmodern cultural condition is Cohen's recent justification of contrived attractions in tourism. According to him, postmodern tourists have become less concerned with the authenticity of the original (Cohen 1995:16). Two reasons can be identified. First, if the cultural sanction of the modern tourist has been the "quest for authenticity", then that of the postmodern tourist is a "playful search for enjoyment" or an "aesthetic enjoyment of surfaces" (1995:21). Secondly, the postmodern tourist becomes more sensitive to the impact of tourism upon fragile host communities or tourist sights. Staged authenticity' thus helps protect a fragile toured culture and community from disturbance by acting as a substitute for the original and keeping tourists away from it (1995:17). Moreover, modern technology can make the inauthentic look more authentic. For example, audiotapes of bird-song can be played repeatedly and in the exact frequency desired by park managers (Fjellman 1992). This technology can make recorded bird-song sound more authentic than actual bird-song since the latter is influenced by the uncertainty of when birds are present and when they sing. As McCrone, Morris and Keily put it,

> Authenticity and originality are, above all, matters of technique.…. What is interesting to post-modernists about heritage is that reality depends on

how convincing the presentation is, how well the "staged authenticity" works. . . . The more "authentic" the representation, the more "real" it is (1995:46).

Thus, the quest for "genuine fakes" (Brown 1996) or inauthenticity is justifiable in postmodern conditions. In Ritzer and Liska's terms,

> Accustomed to the simulated dining experience at MacDonald's, the tourist is generally not apt to want to scrabble for food at the campfire, or to survive on nuts and berries picked on a walk through the woods. The latter may be "authentic", but they are awfully difficult, uncomfortable, and unpredictable in comparison to a meal at a local fast-food restaurant or in the dining room of a hotel that is part of an international chain. Most products of a post-modern world might be willing to eat at the campfire, as long as it is a simulated one on the lawn of the hotel.
> Thus, we would argue, in contrast to MacCannell, that many tourists today are in search of inauthenticity (1997:107).

Both constructivists and postmodernists reveal the crisis of the authenticity of the original (objective authenticity). However, postmodernists are much more radical than constructivists. If constructivists are reluctant to dig a grave for "authenticity" and try to rescue the term by revising its meanings, then postmodernists are quite happy to do so—they have buried it. Indeed, with accelerating globalization under postmodern conditions it is increasingly difficult for the authenticity of the original, such as a marginal ethnic culture, to remain immutable. For postmodernists, gone is the authenticity of the original. Thus, it is no small wonder that they abandon the concept of authenticity altogether and instead justify inauthenticity in tourist space. However, a postmodernist deconstruction of the authenticity of the original implicitly paves the way for defining existential authenticity as an alternative authentic experience in tourism, despite the fact that postmodernists themselves refuse to explore this possibility.

Conceptualizing Existential Authenticity in Tourist Experiences

There has been a long tradition of ontological conceptualization of existential authenticity (Berger 1973; Berman 1970; Golomb 1995; Heiddeger 1962; Taylor 1991; Trilling 1972), ranging from Kierkegaard, Nietzsche, Heidegger, and Sartre to Camus (see Golomb 1995). Existential authenticity has also been a long-term political concern, dating back to the time of Montesquieu and Rousseau (Berman 1970; Trilling 1972). In commonsense terms, existential authenticity denotes a special existential state of Being in which individuals are true to themselves, one which acts as a counterbalance to the loss of "true self" in public roles and public spheres in modern Western society

(Berger 1973). According to Heidegger (1962), to ask about the meaning of Being is to look for the meaning of authenticity. Indeed, there are a number of researchers have discussed the relevance of such an existential authenticity to tourist experiences. For example, Turner and Manning criticize the view that "authenticity is a thing-like social fact, at once a property or characteristic of both actors and settings" (1988: 137). They explain:

> authenticity is only possible once the taken-for-granted world and the security it offers are called into question. This is dependent on a specific mood—anxiety—which, in subjecting everydayness to questioning, reveals the groundlessness of human existence (1988:137).

Turner and Manning clearly show the suitability of applying existential philosophers' (such as Heidegger's) ontological notion of authenticity to tourist experiences. However, they fail to take any further steps towards developing it. After questioning the validity of the conventional concept of authenticity, Hughes suggests that "authenticity must be rethought", and that "one must turn to a qualified existential perspective to recover authenticity in late modernism" (1995:790,796). Neumann hints at an existential authenticity in tourism in his case study of tourist experiences in the Cannon Valley in the United States.

> Travel often provides situations and contexts where people confront alternative possibilities for belonging to the world and others that differ from everyday life. Indeed, part of the promise of travel is to live and know the self in other ways (1992:183).

As previously mentioned, Selwyn (1996a) draws a groundbreaking distinction between "hot authenticity" and "cool authenticity". His concept of hot authenticity, particularly the hot authenticity in relation to myths of the authentic self, is a specific expression of existential authenticity. This realization becomes more evident when he refers to authenticity as the "alienation-smashing feeling". Similarly, what Brown (1996) calls an "authentically ... hedonistic ... good time" illustrates the temporal nature of existential authenticity.

Thus, existential authenticity, unlike object-related authenticity, often has nothing to do with the issue of whether or not toured objects are authentic. In search of experiences which are existentially authentic, tourists are preoccupied with an *existential state of Being*, activated by certain touristic pursuits. To put it another way, existential authenticity in tourism is the authenticity of Being which, as a potential, is to be subjectively or intersubjectively experienced by tourists as the process of tourism unfolds. Daniel's (1996) discussion of "experiential authenticity" in dance performances can be used to exemplify existential authenticity in touristic experiences. Daniel argues that the experiential authenticity linked to dance performances, such as the rumba in Cuba, is derived from tourists participating in the dance rather than being merely spectators. She writes:

> Many tourists are drawn into participation by the amiable feelings, sociability, and the musical and kinesthetic elements of dance performance. Often, not knowing the rules, they do not wait to be invited to dance, but spontaneously join in. They explore their rhythmic, harmonic, and physical potential and arrive at sensations of well being, pleasure, joy, or fun, and at times, frustration as well.
>
> As tourists associate these sentiments with dancing, the dance performance transforms their reality. For many tourists, the dance becomes their entire world at that particular moment. Time and tensions are suspended. The discrepancies of the real world are postponed. As performing dancers, tourists access the magical world of liminality which offers spiritual and aesthetic nourishment. Tourism, in moments of dance performance, opens the door to a liminal world that gives relief from day-to-day, ordinary tensions, and, for Cuban dancers and dancing tourist particularly, permits indulgence in near-ecstatic experiences (1996:789).

Here, if the rumba is treated only as a toured object (spectacle), then it involves objective authenticity in MacCannell's sense, that is, its authenticity lies in the fact of whether or not it is a genuine *re-enactment* of the traditional rumba. However, once it is turned into a kind of tourist activity it constitutes an alternative source of authenticity, i.e., existential authenticity, which has nothing to do with the issue of whether this particular performance is an exact re-enactment of the traditional dance. In reality, as Daniel soon discovers, new elements, are always being integrated into the old rumba. Thus, even though the rumba in which tourists participate may be inauthentic or contrived in MacCannell's sense, it generates a sense of existential authenticity due to its creative and "near-ecstatic" nature.

However, an unanswered question arises with regard to existential authenticity. As mentioned above, existential authenticity in its common-sense acceptance means that "one is true to oneself". This interpretation may seem a little odd at first, since "being true or false" is usually an epistemological issue, a criterion used to judge the nature of utterances, statements, theories, or knowledge. How can the self also be related to the question of "being true or false"? Surely the justification cannot be made in epistemological terms. Rather, one can make sense of the quest for an authentic self only in terms of the *ideal* of authenticity to be found within modern societies. This ideal is formulated in response to the ambivalence of the existential conditions of modernity. It emerges as a reaction to "the disintegration of sincerity"or pretense, and its occurrence is closely related to the feeling of a loss of "real self" in public roles (Berger 1973:82). The ideal of authenticity can be characterized by either nostalgia or romanticism. It is nostalgic because it idealizes the ways of life in which people are supposed to be freer, more innocent, more spontaneous, purer, and truer to themselves than present generations (such ways of life are usually presumed to exist in the past or in childhood). People are nostalgic about these ways of life because they want to relive them in the form of tourism at least temporarily, empathically, and symbolically. It is also romantic because it emphasizes naturalness, sentiments, and feelings in response to the increasing self-constraints of reason and rationality in modernity. Therefore, in contrast to everyday roles, the tourist role is linked to this ideal of authenticity. Tourism is thus regarded as a simpler, freer, more spontaneous, more authentic, or less serious, less utilitarian, and romantic lifestyle which enables tourists to keep a distance from, or transcend, their daily lives. Examples of related pursuits include camping, picnicking, making camp-fires, mountaineering, walk-abouts, wilderness solitude, and adventure. In these activities people are not literally concerned about the authenticity of toured objects. Rather, they are seeking their authentic selves with the aid of tourist activities or toured objects.

However, some may argue that tourism is also subject to constraints (such as the constraint of schedules, itineraries, queuing, finances, etc.) and that the social control exerted by the tourism industry precludes the so-called freedom to be had from tourism, rendering it only a fantasy and illusion. Thus, the question might be asked: isn't existential authenticity in tourism illusory, and hence inaccessible in reality? The point at issue is that an "emic" perspective, rather than an external one is more appropriate when answering this question. Certainly, the tourist experience involves its own constraints. However, such constraints may be seen by tourists as the necessary cost of authentic experiences, rather than as an obstacle to existential authenticity. Indeed, in tourism, existential authenticity may be a fantasy. However, such a fantasy is a real one—it is a fantastic feeling, a subjective (or intersubjective) feeling, which is *real* and accessible to the tourist through tourism. This fantastic feeling is the very feeling characterizing existential authenticity (Dann 1976).

A sense of "authentic self" involves a balance between two parts of Being—reason and emotion, self-constraint and spontaneity, Logos and Eros, or what Freud calls the "reality principle" and the "pleasure principle" (*Chapter 2*). At the risk of oversimplification, to live a life in terms of the dictates of emotions, feelings, spontaneity, or Eros, rather than reason or self-constraints, may be characteristic of a relatively large part of primitive, or precivilized, forms of life. Freud argues that the opposite is the case of civilized or modern forms of life. However, a sense of *in*authentic self arises when the balance between these two parts of Being is broken down in such a way that rational factors over-control non-rational factors (emotion, bodily feeling, and spontaneity, etc.) and leave too little space for satisfaction of the latter. This is the situation characterizing the ambivalence of the mainstream institutional realms of modernity, in which the factors of Logos reign and the factors of Eros are more or less constrained (*Chapter 2*). For example, Hochschild's (1983) empirical study of how American flight attendants are "forced" to present a smiling face to customers illustrates how these attendants lose their authentic selves in the service industry.

Thus, under the condition of modernity, the "authentic self" emerges as an ideal that acts to resist or invert the dominant rational order of the dominant institutions in modernity. To resist the inauthenticity stemming from the mainstream order of modernity, the "authentic self" is often thought to be more easily realized or fulfilled in the space outside the dominant institutions, a space with its own cultural and symbolic boundaries which demarcate the profane from the sacred (Graburn 1989), responsibilities from freedom, work from leisure, and the inauthentic public role from the authentic self. As a result, nature, for example, is seen as typical of such a space. Tourism, and nature tourism in particular, is thus an effective way to promote the search for the "authentic self". Of course, such an "authentic self" is only achieved in relative terms. It is experienced only within a "liminal zone" (Graburn 1989; V. Turner 1973). In such a liminal zone, persons keep a distance from social constraints (prescriptions, obligations, work ethic, etc.) and invert, suspend, or alter routine order and norms (Gottlieb 1982; Lett 1983; Shields 1991). However, in so doing they do not go as far as to abandon Logos (reason), social order, and social responsibilities altogether; moreover, they are ready to return and adapt to the home society again.

Contextualizing the Issue of Existential Authenticity

The possibilities of tourism as a form of existentially authentic experience have been explored in above section. However, the discussion of authenticity-seeking is limited and does not consider the wider context where authenticity arises as an issue. Therefore, in this section, touristic concern for authenticity is explained in relation to the wider context of modernity as a complex system of contemporary institutions.

Ideal-typically speaking, concern about existential authenticity usually only emerges in modernity (Berger 1973; Berman 1970). "In a closed, static society governed by fixed norms and traditions which are accepted by all its members, authenticity has no place in the vocabulary of human ideals" (Berman 1970:xvii). True, traditonal society can also be constraining, but the nature of the constraints is different. By contrast, existential authenticity becomes an issue in modernity because it disappears from the modern structural, public, or impersonal domain in which, ideal-typically speaking, the traditional harmony between individual and society can no longer be maintained. Thus, Berger claims, "The opposition between self and society has now reached its maximum. The concept of authenticity is one way of articulating this experience" (1973: 88). In Sack's terms, authenticity is "a relative evaluation of modern life and place and stands as an indication of our reaction to modern life" (1992:172).

The emergence of the ideal of authenticity can be understood in terms of institutional modernity. In other words, it is the inauthenticity and alienating conditions of institutional modernity that are responsible for the sociogenesis of authenticity as a modern value and concern. In this respect, Marx's critique of alienating labor in capitalism is quite telling. As far as is known, Marx did not use the term "authenticity" in his works. However, he, more than any one else, offers more inspiration for conceptualizing "authenticity" in a broader ontological sense. In his *Economic and Philosophical Manuscripts of 1844* (1977), Marx outlines the ontological conditions of inauthenticity from which a broader range of ontological meanings of authenticity can be derived. The most inauthentic human activities under capitalism are those experienced by "alienated labor". As Marx writes:

> First, the fact that labor is external to the worker, i.e., it does not belong to his intrinsic nature; that in his work, therefore, he does not affirm himself but denies himself, does not feel content but unhappy, does not develop freely his physical and mental energy but mortifies his body and ruins his mind. The worker therefore only feels himself outside his work, and in his work feels outside himself. He feels at home when he is not working, and when he is working he does not feel at home. His labor is therefore not voluntary, but coerced; it is forced labor. It is therefore not the satisfaction of a need; it is merely a *means* to satisfy needs external to it.... External labor, labor in which man alienates himself, is a labor of self-sacrifice, of mortification. Lastly, the external character of labor for the worker appears in the fact that it is not his own, but someone else's (1977:65–66).

According to Marx, ideal-typically speaking, labor should be a free and voluntary activity which is no longer merely a means to ensure survival, but a primary human need. In such an idealized conception of labor, laborers are full "authors" of their labor and thus true to themselves. However, alienated labor under capitalism is an antithesis to this ideal labor (authentic Being). First, it is forced and coerced labor, rather than voluntary labor. Second, it is an instrumental activity, merely a means of ensuring survival, rather than an end or the satisfaction of an intrinsic need. Third, laborers feel unhappy in alienated labor; it is the self-sacrifice and mortification of both their physical and mental well-being. Fourth, it also implies alienation in terms of the interrelationships between human beings, because workers' labor is exploited by their employers; it is not their own. Workers own neither the "authorship" of their work (i.e., their work is designed by the employers) nor the end-product of their labor (which is owned by the employers). Alienated labor is therefore ontologically inauthentic.

Logically, ontological (or existential) authenticity is the opposite of ontological inauthenticity. If Marx does not explicitly define what authenticity is, the meaning of ontological authenticity can be grasped by reading his *Economic and Philosophical Manuscripts of 1844*. In one place, he mentions "man in entire richness of his being" (1977:96). In another place, he writes of "real individual life" (1977:124). Elsewhere he claims that, "The *rich* human being is simultaneously the human being *in need of* a totality of manifestations of life—the man in whom his own realization exists as an inner necessity, as *need*" (1977:99), and that "man in the entire richness of his being—produces the *rich* man *profoundly endowed with all the senses*" (1977:97).

To put it simply, ontological authenticity is about the "real individual life". But what is the real—and hence authentic—life? According to Marx, real individual life involves the "entire richness of his being". In other words, real individual life involves the comprehensive development and fulfilment of personal potential. For Marx, such an ontological authenticity is impossible under capitalism because it is characterized by "alienated labor"; rather, "real individual life" can only be realized in an ideal society (e.g., communist society) in which the full development of each is the condition of the full development of all, that is to say, the contradiction between the individual and society is abolished. As Marx writes,

> Assume man to be man and his relationship to the world to be a human one: then you can exchange love only for love, trust for trust, etc.. . . Every one of your relations to man and to nature must be *specific expression*, corresponding to the object of your will, of your real individual life (1977:124).

Thus, according to Marx, personal authenticity (real individual life) entails an ontologically authentic community or society (i.e., human relationships) in which the quest for personal authenticity is no longer at odds with the community or society as a whole. Apart from that, personal authenticity also entails an ontologically authentic human relationship to nature. Thus, authentic *inter-human relationships*, authentic *human–nature relationship*, and the *authentic self* are dialectically interdependent. The realization of each is the condition for the realization of the other two. Therefore, Marx's "real individual life" refers to a broader meaning of ontological authenticity, in which touristic authenticity-seeking would only be a specific instance.

Ontological (existential) inauthenticity is a malaise found in modernity. Such ontological inauthenticity has found many expressions in modern social theories. Marx's "alienation", Weber's "iron cage",

Durkheim's "anomie", Simmel's "estrangement" in the money economy, Marcuse's "one-dimensional man", Camus's "absurdity", Habermas's "the colonization of lifeworld", and so on are all examples.

Modernity refers to the new social order of the last two or three centuries. From an institutional perspective, modernity—as embodied in capitalism (Marx), industrialism (Durkheim), and formally rationalized bureaucracy (Weber)—is a structural foundation that is responsible for the loss of existential authenticity. The range of loss is broad, including the loss of authenticity from human relationships (the loss of community), human–nature relationships (the loss of natural nature), and intrapersonal relationships (the loss of self), which can be schematically illustrated (Table 3.2).

Table 3.2 indicates the following. First, the capitalist system of commodity production is characterized by the "relations of production" in which the owners of private capital exploit the non-owners who sell their labor for wages (Marx 1954). This impersonal, contractual, and interest-calculated relationship between the owners of private capital and the non-owners has destroyed the traditional, spontaneous, and authentic community in which, ideal-typically speaking, a kind of "social authenticity" is embedded. In capitalist economic relations which entail the "commoditization of labor power", human relationships are certainly no longer authentic. Thus, authentic human relationships must be sought outside the economic system, in the private sphere, in an emotional community, and so on. Second, formally rationalized organizations (Weber), as exhibited in the state and business sectors, impose strict labor disciplines upon individuals which may be at odds with their "real" wishes or unconscious intentions. To adapt to organizational and environmental pressures, individuals must accept strict self-constraint at the expense of (or at least part of) their spontaneity, impulses,

Table 3.2. Institutional Modernity and Ontological Inauthenticity

Institutional modernity	Dimensions	Ontological *in*authenticity
Capitalism (Marx)	Inter-personal relationships	Instrumentalization and commodification of human relationships
Formally rationalized organizations (Weber)	Intra-personal relationships (reason and emotions)	split of self; self-constraints; self-masking
Industrialism (Durkheim)	Humanity–nature relationships	Technological environments, man estranged from nature, environmental crises

and desires. Thus, the formally rationalized organization may give rise to a feeling of loss of personal authenticity. Third, industrialism, as a form of modern inanimate or technological power over nature, multiplies human productivity (Durkheim). However, in so doing, it brings about "risks" and dangers of its own (Beck 1992). As a result, humans have lost an authentic relationship with nature.

What underlies these three mainstream institutions of modernity (capitalism, industrialism, and the formally rationalized organisation) is the principle of the "primacy of instrumental reason" (Taylor 1991). About instrumental reason, Bauman writes:

> the modern way of acting is described as *rational*: dictated by instrumental reason, which measures the actual results against the intended end and calculates the expenditure of resources and labor. The catch, however, is that not all costs are included in the calculation—only those that are born by the actors themselves; and not all results are monitored—only those that are relevant to the set task as defined by or for the actors. If, on the other hand, all losses and gains were taken into account ... the superiority of the modern way of doing things would look less certain. It might well transpire that the ultimate outcome of the multitude of partial and separate rational actions is more, not less, overall *irrationality* (Bauman 1990b: 193–194).

Therefore, although the institutional complex of modernity brings about extraordinary material, organizational, and intellectual achievements, it is essentially ambivalent. The institutional complex of modernity supplies people with affluence, freedom, and social order, but also leads to the loss of authenticity in the dimension of their relation to others, to nature, and to themselves.

However, to say that modernity leads to ontological inauthenticity should not imply that modernity is totally evil. On the contrary, given that any society is both "evil" and "good", or enabling as well as constraining in Giddens's (1979) sense, modernity, especially late modernity, is the most enabling and liberating institutional order that human beings have found so far, although it is also constraining. Self-evidently, complaints against the dark side of modernity are made on the basis of a basic satisfaction with the living conditions altered by modernity. People from the Third World emigrate to advanced societies because they want to have immediate access or a short-cut to the condition of modernity, which is thought of as better than the conditions found in their own countries. By contrast, people in developed countries are eventually satisfied with their own societies (Bruner 1991:240). Therefore, it is more correct to say that modernity is ambivalent. It is this ambivalence at the heart of modernity that tourism has to confront.

Modernity, especially late modernity, has triumphed because most people prefer it to a traditional form of life. "The achievements of modernity in human liberation and culture are obvious even though it is fashionable today to emphasise the bad results of many modern strategies" (Kolb 1986:260−261). However, as modernity progresses its discontents grow as well, partly because of its existential inauthenticity. The emergence of such discontents implies that something needs to be changed. Obviously, if one cannot expect a total change of the project of modernity as a whole, then the immediate and simplest solution is to leave the home environment. "For someone who dreams of changing lives, of changing life, travel is the simplest approach" (Todorov 1993:271). Thus, modern people like travel. Of course, the concrete purposes of travel are various and innumerable, but in a most common meaning travel has been implicitly adopted as an action that *resists* the condition of existential *in*authenticity.

Arising from this resistance to the ontological inauthenticity of modernity is a sense of existential authenticity. Modern people travel away from home in order to "forget" or "escape" the malaise of modernity by entering an alternative, fantastic, and separate world. Tourism is one way of accessing existential authenticity since it resists the logic and ethic of everyday reality and offers an intensified and concentrated experience of an alternative Being-in-the-world. If daily reality is characterized by complexity, artificiality, and self-constraint, in short, ontological inauthenticity, then tourism provides access to a socially and culturally constructed "utopian" world in which people legitimately experience simplicity, naturalness, and "communitas" (Turner 1973). Although tourism is about a "utopian" or "fantastic" world, it is a *make-believe* world. If tourists think that they have achieved a sense of personal, interpersonal, and human-nature authenticity, then their feelings are ontologically real to themselves. How tourists seek and experience their personal and interpersonal authenticity will be discussed in turn below. The experience of authentic human−nature relationships in tourism will be discussed in the next chapter in relation to the technological environment of modernity.

Concretizing Existential Authenticity in Tourism Experiences

Analytically speaking, existential authenticity can be divided into two different dimensions: *intra*-personal authenticity and *inter*-personal authenticity. Both can be achieved by means of tourism. As previously indicated, nature, as a space signifying freedom from the structural constraints of society, is most often toured and used as a medium to help learning about a sense of authentic self or intersubjective authenticity.

Intra-personal Authenticity in Touristic Experiences

Bodily Feelings. Obviously, the intra-personal dimension of existential authenticity involves bodily feelings. The body or concern for the body has recently attracted wide academic attention, partly as a reaction to the dominance and longevity of the Cartesian–Kantian tradition which enhances the status of mind at the expense of body. Concern for the body is also thought of as an important aspect of tourism (Veijola and Jokinen 1994). Relaxation, rehabilitation, diversion, recreation, entertainment, refreshment, sensation-seeking, sensual pleasures, excitement, play, and so on are all bodily experiences (Cohen 1979b, 1985; Lett 1983; Mergen 1986). Touristic search for bodily pleasure also exhibits the characteristics of a ritual, that is, the recreation ritual (Graburn 1983a:15). Roughly speaking, concern for the body has of two aspects: the sensual and the symbolic. Whereas the latter comprises the culture or sign-system of the body (Featherstone et al 1991), the former involves bodily feelings. On the one hand, in relation to the culture of the body, the body becomes a "display" of personal identity (health, naturalness, youth, vigor, vitality, fitness, movement, beauty, energy, leisure class, taste, distinction, romance, etc.) (Bourdieu 1984; Featherstone 1991a; Rojek 1993). On the other hand, the body is the primary organ of sensibility or feeling. Thus, the body is the inner source of feelings and sensual pleasure. In this sense the body is not merely a corporate substance but also a "body-subject" or the "feeling-subject" (Seaman 1979).

The body is a battlefield. Control and manipulation of the body gives rise to power (Foucault 1977). Part of the power that modernity has over the body comes from the surveillance of a population (Giddens 1990). Another aspect of this derives from time-space structures relating to work and the division of labor (Lefebvre 1991). The commodification of labor power entails the disciplines of labor and the regular presence of the body (the bearer of labor power) in certain structured spatio-temporal areas (workdays and workspace). In both situations, self-control of bodily drives and impulses is necessary.

The power derived from control over the body in the latter sense results in an experience of existential *in*authenticity. In other words, existential inauthenticity or alienation is not only spiritual but also bodily. Therefore, concern with bodily feeling is in fact concern with the bodily, or intra-personal, source of the "authentic self". There is no better place than the beach to illustrate bodily concern with the authentic self. On the one hand, in this setting the body shows that it is relaxed and not limited by bodily control or self-control imposed by social structures or the superego. On the other hand, the body alters its routine existence and enters an alternative, yet intensified, experiential state: recreation, diversion, entertainment, spontaneity, playfulness, in short authenticity in the exis-

native source of authentic experiences in tourism. In a number of kinds of tourism such as nature tourism, landscape tourism, green tourism, holidays on the beach, ocean cruising, hobby tourism, adventures, family vacations, visiting friends and relatives, and so on, what tourists seek are their own authentic selves or inter-subjective authenticity, and the issue of whether toured objects are authentic is irrelevant, or less relevant. As the category of existential authenticity can explain a wider spectrum of tourist phenomena than the conventional concept of authenticity, it therefore opens up broad prospects for the rejustification of "authenticity-seeking" as the foundation of tourist motivation. Furthermore, authenticity-seeking in tourism is explained sociologically in relation to the wider context of modernity in which authenticity arises as a modern value and concern. Tourism is thus one of forms of quest for authenticity in response to the structural inauthenticity of modernity.

In this chapter, the limits of objective and constructive authenticities have been exposed. However, their relevance to tourism is not negated altogether. Further efforts need to be made to discover the empirical relationships between objective, constructive (or symbolic), and existential authenticities, the extent to which each of these authenticities is the major concern of tourists, and the reasons why certain tourists prefer one kind of authenticity over others.

From an economic perspective, tourism is, in a sense, an industry of authenticity. Tourism involves both the supply and consumption of the commodity of authentic experiences. Existential authenticity becomes a commodity, or a commoditized experience, only in the context of modernity. Existential authenticity' is thus commercially transformed into the packaged experience of "sun, sand, surf, and sex", of the Garden of Eden, the idyllic rustic life, and so on. All these offerings are culturally sanctioned and socially constructed. Existential authenticity is indeed an implicit selling point of the products of tourism. In this sense, tourism is a "dream industry", and buying a holiday is buying a chance to have a dream come true. Thus, metaphorically speaking, modernity uses one hand to take people to existentially inauthentic situations, but at the same time it uses the other hand to show pictures of a dream which promises people "salvation" and, as a result, keeps them in its tracks.

Chapter 4

Modernity and Nature Tourism

Technology is not a new and distinctive phenomenon in modernity; previous civilizations also had technologies. What is new is that in the context of modernity technologies have been "projected and embodied in organized forms which dominate every aspect of our existence" (Mumford 1934:4). Technological advance is, so to speak, one of the most convincing and striking dimensions of modernity. Marx's famous dictum "All that is solid melts into air" describes the consequences brought about by modern technological advance. One such effect is the dramatically increased mobility of the modern subject (Lash and Urry 1994; Urry 1995). The connection between modern technologies and modern tourism is obvious enough, but it can often be overstated, as Urry (1995) notes, as a form of technological reductionism. In many textbooks on tourism it is taken for granted that modern tourism is simply the product of technological advances, such as those in transportation and communication, as well as increased personal disposable income and time. Tourism is seen as the *immediate* and *direct* extension and application of transportation and other technologies into leisure areas as part of a general improvement in living standards. The tourist is seen as a rational actor who is ready to take full advantage of the latest advances in technology.

Such a model is useful and correct to some extent, but it suffers from several limitations. First, it ignores the role played by organizational innovation (such as that pioneered Thomas Cook), which is no less important than new transportation technologies. Tourism is not simply the immediate extension of new technologies into the field of travel, it also involves the organizational transformation and viability of new technologies necessary to economic success (Urry 1995:142). Second, it ignores changing preferences and tastes (Johnson and Thomas 1992:3). If transportation technologies provide the physical precondition for new patterns of travel, then it is the changing preferences and tastes that make a particular pattern of travel culturally acceptable. Third, it fails to consider negative cultural reactions to some technological advances. Historically, there used to be hostility towards railway travel for the purpose of sightseeing (Schivelbusch 1986). Indeed, the romantic movement, which originated in the late 18th century, was a negative cultural reaction to the age of the machine (Mumford 1934). Technologies are, in short, not necessarily welcomed. Finally, it ignores the significance of the *cultural* construction of travel and tourism. Tourism is not exclusively the product of technological advance and economical improvement (Urry 1995:142), but it is also the outcome of cultural change (Andrews 1989; Green 1990; Jasen 1991; Ousby 1990; Squire 1988; Urry 1990a). For instance, in the heyday of romanticism some tourist activities were in fact the result of a cultural, romantic, or nostalgic critique of industrial and technological civilization. This situation was manifested in an enthusiastic search for picturesque and sublime landscapes unspoiled by technological civilization (Andrews 1989; Jasen 1991; Ousby 1990; Squire 1988).

Such technological (economic) reductionism of tourism finds its psychological counterpart in the Maslowian psychology of tourism. According to Maslow, there is a hierarchy of needs and individuals do not strive to satisfy all these needs simultaneously, rather the more derived needs—self-expression, spiritual fulfilment, and so on—are satisfied only after the more basic needs—food, shelter, and procreation—have been met. For its part, tourism is possible only when more basic needs have been satisfied. This view implies that tourism is only possible when there is sufficient disposable income and time, as well as efficient transportation and other relevant technologies. Such a proposition is only partialy true. One of the problems this suggestion faces is the fact that a hierarchy of needs is culturally specific; it is relative rather than universal. A need defined by one culture as basic may become a derived one in another culture and vice versa. Western aid providers in Nepal, for example, have been horrified to find that poor villagers spend their money on refurbishing their temple rather than improving the agricultural conditions of their rice fields. For Western aid providers an adequate supply of rice is considered to be the most basic need of the villagers, but the latter do not think this way. Establishing a good relationship with their gods is thought of as no less basic a need than the supply of rice (Thompson, Ellis and Wildavsky 1990:55).

Certainly the rise of tourism is dependent on the tourist having "a surplus over bare subsistence" (Pimlott 1976). However, in terms of Maslowian psychology it is difficult to explain why a person is necessarily willing to spend this surplus on tourism rather than something else. The cause of the rise of modern tourism is not merely technological or economical, nor is it solely psychological; it is also *cultural*. Modern tourism involves the emergence of a new culturally made subjectivity. Therefore, in addition to other causal forces, "once the self is operational, it becomes an independent causal agent" (Csikszentmihalyi 1988:17). The gap between technological conditions and the genesis of tourist activity is filled by cultural factors. Therefore, it is crucial to identify the cultural genesis of modern tourism. This does not imply any denial of the important roles played by technological advances and the related improvement of living standards. Rather, it helps overcome any rude and simplistic technological/economic reductionism.

Technology can be double-edged: it can be either the means of death and destruction (e.g., nuclear bomber, bioweapons, technological accidents) or it can be the means of protection and enablement. In relation to tourism, technology is also perceived to be highly contradictory and ambivalent: it can be positive by enabling travel, such as transportation technology (e.g., the growing use of jets in the post-war period; thus tourism as consuming and celebrating technologies), or it can be negative, appearing as a constraining and depressing daily environment from which people really want to escape. Tourism is thus in a sense a negative reaction to and an attempt to escape from the technological environment. Each of these aspects of technology—the enabling aspect, characterized by consumption and celebration of the technology, and the constraining aspect, where people are pushed away—are *mediated by cultural factors*.

Technologies do not automatically determine modern tourism since they are culturally mediated. Tourism is linked to technologies, but the relationship between them is not merely a technological or economic one; it is also cultural. The cultural relationship between tourism and technology is multidimensional; it can be either positive (thus celebrating) or negative (thus escaping). In the early stage of modernity, tourism partly originated as a negative and hostile reaction to the industrialized urban environments that were associated with technological change and civilization. The case of tourism in the Lake District, shaped by the romantics in the late 18th century, is one example (Squire 1988; Urry 1995). The process whereby technologies came to be culturally assimilated and accepted was gradual (Mumford 1934).

Hence, there are basically two different ideal-typical cultural orientations towards, and cultural transformations of, technological civilization with which modern tourism is connected: the idealistic/romantic orientation, and the materialistic/hedonistic orientation. The former tends to focus on the negative/depressing consequences of technological civilization, particularly the urban environment, built with the modern technologies associated with industrialization. Conversely, the latter focuses on the positive/liberating aspects of technology. In short, from a cultural perspective, technological civilization is *ambivalent*. This situation has of course been echoed in the culture of tourism.

It is acknowledged here that technological advances, such as those in transportation and communication, are extremely important to modern and contemporary tourism (especially mass tourism), but, as the positive functions of technology in tourism are well documented they will not be repeated here. Rather, this chapter focuses on the *idealistic/romantic* critique of the negative/depressing features of technological or urban physical environments, and on the impact on tourism of negative cultural reactions to technological environments. The chapter consists of three sections. In the first section, the relationship between modernity and technology is outlined. The second section examines the romantic roots of modern tourism. The third section discusses the interaction between romanticism and the technological or urban physical environment.

Modernity, Technology and the Environment

Nature is the original home of humankind. Nature is also dangerous and represents various hazards, threats, and disasters. Therefore, the "drive to modify the natural or given environment so that it will be safer, more abundant, and more pleasing is as old as humankind" (Williams 1990:1). It is during humankind's transformation of nature that technologies and artificial environments such as houses, villages, towns, and cities gradually appear. In this sense, "Environment and technology form not a dichotomy but a continuum", and the "human environment has always been, to some degree, artificial" (1990:1).

But the "degree of the artificiality is what has changed so radically in modern times" (1990:1). Modernity, to a degree, originates from industrialization, involving subduing and conquering nature through scientific and technological means. The physical dimension of modernity consists of systematic technological systems, manufactured goods, and artificial or built environments. In the condition of modernity, "nature literally ceases to exist as naturally occurring events become more and more pulled into systems determined by socialized influences", and people thus increasingly "live in artificial environments"; nature is now "represented only in the form of the 'countryside' or 'wilderness'" (Giddens 1991:166).

Modern technological control and conquest of nature is epitomized by the modern (particularly early modern) philosophy of nature, as exemplified in the works of Bacon and Descartes. During the period of modernity, which was characterized by instrumental reason, nature came to be dominated, disenchanted, reduced to the status of what Heidegger (1978) calls "the standing-reserve" of technology, and treated "as instrumental, the means to realise human purposes" (Giddens 1991:165). "Nature was seen as waiting to be 'mastered' and many of Bacon's writings emphasized the way in which male science could and should dominate female nature" (Lash and Urry 1994:293).

Modern Technologies and the Built Environment

Under the condition of modernity technologies penetrate every aspect of life and they are, according to Williams,

> best considered as environments rather than as objects.... From such an environmental perspective, technological change is best evaluated in terms of the general direction of change rather than in terms of the supposed effects of this or that device (1990:127).

Thus, as Tester puts it, "technology tends to come to take on that overwhelming and independent existence which had once been attributed to God or to nature" (1993:85). Indeed, technology, in the context of modernity, has replaced God and become a new idol to worship.

> It is prone to enchantment virtually in direct proportion to the disenchantment of nature . . .the destruction of natural artifice meant at least in part the construction of technological artifice (1993:89).

Nature is no longer the masterpiece of God, but rather something that is amenable to human manipulation and control. God is dead, as Nietzsche claimed; and God was in a sense killed by technology in the modern world.

Technology, the product of modern rational subject, has thus been reified in the condition of modernity (Tester 1993:85,100). Just as Marx described "commodity fetishism", modernity has also witnessed the emergence of technology fetishism. What is produced by humans then confronts humans as "an independent system in its own right" (1993:85). In other words, "technology . . . has become the second natural artifice" (1993:101). From the perspective of reification, technology is today the "second physical environment" or "second nature". Mumford calls this second environment "megatechnics":

> In terms of the currently accepted picture of the relation of man to technics, our age is passing from the primeval state of man, marked by the invention of tools and weapons, to a radically different condition, in which he will not only have conquered nature but detached himself completely from the organic habitat. With this new megatechnics, he will create a uniform, all-

enveloping structure, designed for automatic operation (Mumford 1966:303).

In addition to Mumford's "megatechnics", several other terms are used to describe the second physical environment: the "technological environment" (Williams 1990), "built space" (Giddens 1981), "manufactured environment" (Mumford 1934), "artificial technological environment" (Tester 1993), and so on. Moreover, efforts have been made to construct a model of this modern technological environment. What is the most typical model of the technological or built environment? For some it is the city (Giddens 1981); for others it is a spaceship or an underground world (Williams 1990), and so on. Williams describes the underground in the following:

> Subterranean surroundings, whether real or imaginary, furnish a model of an artificial environment from which nature has been effectively banished. Human beings who live underground must use mechanical devices to provide the necessities of life: food, light, even air. Nature provides only space. The underworld setting therefore takes to an extreme the displacement of the natural environment by a technological one. It hypothesizes human life in a manufactured world (1990:4).

Technologies are also responsible for the dramatic expansion of the built environment such as the urban environments of modern times. The city, or built space, can be traced back to ancient Greece and China. However, the intensive involvement of technologies in the city is a unique feature of modernity. In modern times, cities as built environments are contingent on technology in at least two senses. First, the emergence of numerous industrial cities in the period of modernity is related to industrialization, which involves systematic and institutional application of technologies in the production process. Second, throughout this period cities have seen new technologies put to use for various architectural and administrative purposes (e.g., underground technologies, supply of running water, sewage disposal, traffic administration). In Heidegger's (1978) terms, technology is about the building of dwellings. And it could be said that the most magnificent technological achievement in terms of the building of dwellings is the city. In short, cities become built technological environments which move further and further away from nature.

Technology and Environmental Quality

Tourism can be related to technology in at least two senses. First, from a *positive* angle, technologies, especially transportation technologies, provide the most efficient and comfortable way of getting to, staying in, and returning from tourist destinations. They also increase the opportunities for tourists to travel further afield, which is well illustrated by the example of the role played by jet aircraft in the mass tourism of the post-war age. In other words, modern technologies can enable people to travel; and the consumption and experience of these technologies is part of a tourist's experiences. This aspect of the relationship between tourism and technology is so obvious that there is no need to discuss it in detail here.

Second, from a *negative* angle, technologies, particularly technological environments (e.g., factories) and built environments (e.g., urban settings), are in a sense the forces that push urban dwellers away from artificial environments to experience natural environments. In other words, technologies have to certain extent led to a deterioration in the quality of the environment, which motivates people to travel in search of more pleasing natural environments in order to improve their spiritual and physical well-being. The following discussion concentrates on the second aspect of the relationship between tourism and technology.

One of the essential aims of technology is to protect humankind from the severity, dangers, and threats that result from nature. In broader terms, it is for the purpose of improving the physical environment that humans develop technological civilization. As Tester states:

> The modern technology represented so many practical and material attempts to ensure the possibility of the building of a magnificent and self-sufficient dwelling in the world. It stepped in to the gap which appeared when the order of things could no longer be assumed to be constructed by nature (1993: 84–5).

However, things often go in the opposite direction from that intended. Technology is designed to improve human environments, but due to "the primacy of instrumental reason" that is characteristic of modernity (Taylor 1991), technology frequently creates an "environmental crisis": a crisis in both the physical and the cultural (or phenomenological) sense. Technological rationality ensures the most economical application of means (technologies) to realize a given end. Maximum efficiency is the measure of its success (1991:5). However, each given end is often achieved at the cost of a higher, more comprehensive, or ultimate end—overall environmental quality (an aspect of the quality of life), including not only physical but also psychological and social quality. Some technologies, particularly those which are environmentally unfriendly, may be evaluated as rational according to some partial or organizational criteria and goals, but may at one and the same time be irrational if evaluated from larger societal perspectives and in terms of more substantial criteria, for they may bring about irrational and negative consequences, such as long-term environmental damage. The capitalistic nature of running such technologies undoubtedly encourages such short-termism and the "primacy of instrumental reason". Therefore, in the condition of modernity, evaluations of environmental quality are paradoxical. On the one hand, the physical conditions of modern technological environments may appear better, especially the protection and support that they provide. On the other hand, the phenomenological dimension of these environments may be worse than before because of the undesirable consequences of a deterioration in the quality of the environment.

Environmental quality is thus an indispensable part of the quality of life as a whole. This includes not only the physical quality of the environment but also its phenomenological, experiential, affective, or aesthetic quality (Buttimer and Seamon 1980; Nasar 1988; Russell 1988; Seamon 1979; Tuan 1974, 1977). Or in Williams' terms, environmental quality involves physical, psychological, and social aspects (Williams 1990:2). Under some circumstances, technological and urban environments contain the physical capacity and an adequate material base to support and protect life in a better way, but may simultaneously cause unquantifiable psychological and social discomforts (1990: 2). Of these Williams writes:

> Our environment will inevitably become less natural; the question is whether it will also become less human. The human environment is by definition technological to some degree. But if we allow technology to take over our surroundings, they can become inhospitable to human life (1990:213).

It is clear enough that deterioration of the physical environment brings about psychological discomfort. People are horrified by the numerous environmental disasters caused by technologies, such as leakage from nuclear power stations, chronic environmental pollution, the possible future disappearance of the ozone layer and rain-forests, the number of animal species on the decrease, and so on. In other words, late modernity is increasingly becoming what Beck (1992) calls a "risk society" to which all human beings are exposed. Even these physical environments which protect and support life do not comfort people entirely because technological environments are essentially ambivalent. They may provide the physical protection and support necessary for human survival and the maintenance of well-being, but they may also be psychologically depressing. They may additionally cause intolerable and negative psychological feelings that lead to sense of rootlessness and helplessness in highly complicated technological complexes. Hence, they cause a desire among people to search for roots, simplicity, wilderness, and authenticity in natural environments that are relatively unspoiled by artificial technologies.

One key source of these psychological discontents is the disappearance of nature following its domination by technologies. Humankind is *part of nature* (Lash and Urry 1994:293), a human being may have an inborn preference for natural amenities (including visual amenities). "Nature", says Taylor, "draws us because it is in some way attuned to our feelings,... Nature is like a great keyboard on which our highest sentiments are played out. We turn to it, as we might turn to music, to evoke and strengthen the best in us" (1989:297). But whilst providing the material foundations for a better life, modern technologies also destroy such natural amenities and sentiments in everyday life. The "grey jungles" of high apartment buildings in big cities, for instance, offer people comfortable flats in which to live, but simultaneously isolate them from nature and original natural amenities. Modern science and technology, as Giddens suggests, play a fundamental role in "the sequestration of experience", which includes the sequestration of experience of nature (Giddens 1991:8,156).

According to Williams, from a historical perspective there may be two fundamental spiritual breakdowns related to environmental changes. She quotes the historian of religion, Mircea Eliade, to show how the first "spiritual breakdown" took place during the Neolithic shift from a pastoral to an agricultural civilization. A profound spiritual crisis occurred when the nomadic existence gave way to an agricultural way of life, in which humans no longer wandered freely but were bound to the soil (Williams 1990:2). Elaborating on Eliade's view, Williams suggests that there has been comparable spiritual breakdown during the modern period:

> that humanity's decision to *unbind* itself from the soil—not return to a nomadic existence, but to bind itself instead to a predominantly technological environment—has provoked a similar profound spiritual crisis. We are now embarked upon another period of cultural mourning and upheaval, as we look back to a way of life that is ebbing away (1990:2).

It is perhaps this spiritual crisis concerning environmental change that constitutes one of the most important cultural foundations for modern and contemporary tourism, particularly nature tourism: the technological environment is ambivalent, in that it is not only celebrated but also complained about. One of these complaints is the romantic reaction, which has had a considerable impact upon nature tourism.

Romanticism and the Rise of the Natural Landscape

Nature is today universally regarded as a source of pleasure. Before the eighteenth century the story in the West was different. As Squire puts it,

> until the flowering of the romantic ideology in the late 18th century, Western culture tended to see nature very negatively. In contrast to cities ... wilderness was considered chaotic and therefore dangerous (1988: 238).

For example, Wales, a popular tourist destination today, was described by Daniel Defoe in 1724 as a country "full of horror" (Squire 1988:238). Similarly, before the 18th century the Lake District was described as inhospitable. Daniel Defoe described Westmoreland in the Lake District as "the wildest, most barren and frightful of any that I have passed over" (quoted in Urry 1995:193). Indeed the wilderness was for long "a place to fear" in the West. It is no accident that the wilderness (e.g., forests) was seen as the "home of evil spirits" (Short 1991:6–8). As far as the sea was concerned, its attraction and charm were discovered only during the 18th century. Before that, "the classical period knew nothing of the attraction of seaside beaches, the emotion of a bather plunging into the waves, or the pleasures of a stay at the seaside" (Corbin 1994:1). In other words, the sea was a source of horror, associated with the repulsive image of the Great Flood described in the book of Genesis (Corbin 1994:1–18). Or in Seligman's words, "the sea has been feared and often hated" (1950:74).

Thus, even if human beings may have some inborn preference for specific natural amenities, under certain circumstances they may be repressed by a fear of the uncontrollability of nature. "Nature as landscape was, then, a historically specific social and cultural construction" (Urry 1995:175). In Western culture, the rise of the notion of landscape is linked to the rise of romantic taste and the romantic movement that began in the later stages of the 18th century (Andrews 1989; Jasen 1991; Ousby 1990; Squire 1988).

"Romanticism" is a vague term that resists precise definition. There is a range of various and even competing definitions. The romantics themselves, for instance, defined this term quite differently and from various perspectives (Furst 1976). The philosopher Bertrand Russell tentatively linked it to "a way of feeling" (1979:651) and regarded Rousseau as the father of the romantic movement (1979:660). Some radicals adopt a Freudian approach by linking romanticism to the id as a contrast to classism (linked to the superego) and realism (linked to the "reality principle"); a French critic simply insisted that romanticism should be felt but never defined (Riasanovsky 1992:69).

Furst (1976) has traced the evolution of the term "romantic". In the early Middle ages "romance" referred to the new vernacular languages, as opposed to the learned tongue, which at that time was Latin. In Old French, "roman" referred to a courtly romance in verse as well as a popular story. This term first became familiar and widely used in England. At first it was related to the old romances and tales of chivalry, which were characterized by high-flown sentiments, improbability, exaggeration, and unreality. During the Age of Reason in the 17th century the word fell into increasing disrepute and was put alongside "chimerical", "bombastic", "ridiculous", and "childish". It was only from the early 18th century that the term gradually began to recover its status and gained a positive meaning in English society. As early as 1711 the word "romantic" was associated with "fine". Later (from the mid-18th century onwards) it was connected with "imagination" and "feelings". Moreover, it referred to landscapes and scenes (Furst 1976:12–13). Thus, "romantic" was not, from its inception, a term of belonging to artistic criticism but denoted a turn of mind that looked favourably on things of an imaginative and emotional kind. Later on it was applied to the romantic movement in literature (1976:13–14). A taste and passion for landscapes and nature was one of the original characteristics of the romantic movement and romanticism.

Thus, tourism in Western societies can be linked to romanticism in two senses. First, tourism, particularly landscape or nature tourism, originated from "a turn of mind" and the emergence of a romantic taste for natural landscapes amongst the intellectual élite in the later stages of the 18th century. Second, famous romantics such as Wordsworth, through their poetic or romantic discourses on the taste and love of nature, in turn shaped nature tourism. The romantic feelings, emotions, and experiences of nature which were finely described and recorded in such discourses have become what many travelers and tourists now enthusiastically seek when they visit the Lake District, or more generally, nature (Andrews 1989; Buzard 1993; Ousby 1990; Squire 1988).

The Rise of the Romantic Taste for Natural Landscape

"Taste" is originally denoted a preference for particular kinds of food rather than others. The term has been elaborated on in cultural studies (see Bourdieu's (1984) classic study of culture and social judgement of taste). Taste is also crucial to tourism. In his study of the history of culture and nature tourism in England, Ousby has shown "how movements in taste have led to patterns of travel, and how these patterns of travel have in turn been expanded and systematized into a tourist industry" (1990:5).

Tourism, particularly nature tourism, involves a taste for landscape. For a tourist, nature is rarely seen as neutral. In other words a tourist *appreciates* nature rather than simply sees it in a cool, detached, indifferent, or scientific way. The fact that nature is transformed as "landscape" presupposes the formation of new taste, namely, a romantic or aesthetic stance rather than a utilitarian stance towards nature. Zukin puts it this way: "The material landscape was mediated by a process of cultural appropriation, and the history of its creation was subsumed by visual consumption" (1992:225). Such a visual consumption of landscape is possible only when a romantic taste for nature and landscape has occurred. Thus, nature tourism entails romantic taste that refers to

> the application of general tendencies of thought and cultural attitude to the act of judging one aspect of our environment as interesting, beautiful or otherwise worth attention and rejecting other as not. Travel quickly converts these judgements into practical, local and specific terms. In doing so, it creates a habit of vision and a corresponding habit of blindness: seeing our environment, getting to know a region of England or an aspect of its life, increasingly become a matter of appreciating particular sights from a particular angle. Tourism completes the process by turning the habits of travel into a formal codification which exerts mass influence and gains mass acceptance (Ousby 1990:5).

Two related aesthetic principles, the sublime and the picturesque, were integrated as parts of romantic taste for landscape (Jasen 1991:286–287). The picturesque, at its simplest, means "that kind of beauty which would look well in a picture" (Ousby 1990:154). In fact the term "landscape" was originally used by painters to allude to the pictorial representation of scenery. Gradually it was widened to denote the scene itself, viewed pictorially and from a fixed vantage point. Finally it described a whole tract of countryside or rural scenery in visual terms (1990:154). The earlier age's picturesque tourist usually appreciated a landscape by comparing it to a famous painter's paintings or poet's poems. As Andrews claims: "the tourist travelling through the Lake District or North Wales will loudly acclaim the *native* beauties of British landscape by invoking idealized *foreign* models—Roman pastoral poetry or the seventeenth-century paintings of Claude and Salvator Rosa" (1989:3). Later, English picturesque tourists moved away from comparing the landscape to examples of foreign poetry or paintings, and compared it to native poetry and paintings instead. Such "recognition and tracing of resemblance between art and nature" was "one of the chief excitements for the picturesque tourist" (1989:39).

The sublime is another important category in the judgement of taste of landscapes. The term "sublime" was well elaborated by the English philosopher Edmund Burke. In his influential *Philosophical Enquiry into the Origin of Our Ideas of the Sublime and Beautiful* (1757), Burke linked the sublime to one of humanity's strongest instincts, self-preservation. When this instinct is threatened, terror is caused; and terrifying experiences are the source of the sublime. Thus, the experience of the sublime involves excitement with a mixture of awe, terror, and admiration when viewers are confronted by natural phenomena which remind them of their relative insignificance, fragility, or bodily powerlessness compared with the power of such phenomena (e.g., huge mountains, deep ravines, big falling waters, violent thunderstorms, vast oceans, etc.). The gap between the sublime and fear may be very narrow. The shift from fear to sublimity presupposes a self-confidence which is partly derived from guaranteed safety (and partly from cultural cultivation) when facing sublime natural phenomena. The sublime is essentially modern because it represents a taste for those parts of nature that are not yet been fully under control (or have not been controlled because it would not be profitable, such as transforming deserts into fields). Such a taste is associated with the growth of humankind's overall power over nature in general in modernity. It is only against such a background that the sublime can become one of the pleasant experiences that the modern tourist seeks.

Of course, such a taste for landscapes and appreciation of both the picturesque and the sublime, existed only amongst a small circle of educated élites (Squire 1988:243), or as Andrews (1989) calls them, the "men of taste", during the early stage of modernity, particularly during the late 18th century. At that time, not all people could afford the intellectual education in the arts and literature that was required for an appreciation of landscapes (1989:3). However, as time went by, such a taste gradually spread to wider circles of people, particularly to members of the middle class. Even today the taste for landscape expressed by the early romantics is still alive, constantly renewed, and has even become one of central cultural values in the contemporary world, which transcends class conditions and appears, so as to speak, a common taste. Increased environmental consciousness has bestowed new meanings and sentiments on nature (Urry 1992). In short, today's popularity of nature tourism has its romantic roots.

The Influences of Romanticism upon Nature Tourism

The old romantics were the "persons of taste", an intellectual élite who set in motion one of the most influential cultural tendencies during the period of modernization. The romantics' discourses represent what Lash (1993) and Urry (1995) call the "aesthetic reflexivity" of modernity. Such discourses have greatly shaped cultural preferences and practices in the contemporary world. Campbell (1987) even argues that romanticism has had an influential economic impact. According to Campbell, romanticism has helped bring about a new hedonism in which the stimulation of emotion or "feelings" becomes a major source of pleasure. Due to the romantic ethic, objects and images are increasingly associated with feelings, which foster a desire for consumer goods and services. This desire in turn fuels the ever-expanding system of production and perpetual innovation that characterize modern capitalist society.

The romantic movement has also had a considerable influence upon tourism. This influence can be explained in two senses. First, the taste for landscape fostered and shared by the romantics gradually filtered to the wider society, becoming more universal and fuelling people's lust for nature and their love of nature or landscape tourism. Second, the works or discourses of the romantics formed a perceptual framework or a structure of perception through which tourists came to see landscape. The sights mentioned and described by the romantics became sites of worship and pilgrimage for later tourists. They regarded it as a pleasure to gain similar emotional experiences from the same sights and, if possible, to look at the landscapes from the same perspectives as the romantics did (Jasen 1991; Ousby 1990; Squire 1988).

Turning to the second sense, Ousby claims that: "Romanticism had a profound influence on tourism: by changing people's perception of what was admirable or beautiful, it altered and expanded the list of sights people visited on their travels" (1990:99). The most influential romantic figure on tourism is perhaps Wordsworth. Through reading the poetry of Wordsworth, the natural landscape becomes, in a sense, a literal landscape. Ousby describes this phenomenon as such:

> Seeing the landscape—we might now say almost literally "reading" the landscape—in terms of the poems describing it became the nineteenth-century equivalent of the earlier systems that the cults of the Sublime and the Picturesque had established for teaching visitors what to look at, what to see and what to feel. The cultivated tourist convinced himself that he had experienced the distinctive atmosphere of the places he visited, or even come to know their essential character, by opening his volume of Wordsworth (1990:181).

Squire (1988) has also examined the impact of English romanticism upon tourism through a case study of Wordsworth's influence upon Lake District tourism. For Squire, "romantic literature helped to foster public appreciation of wild country and primitivism" (1988:238). "The popularity of romantic literature has also fostered tourism; hordes of visitors, anxious to recreate the emotional experiences in place described by a literary idol, still descend on areas immortalized in poetry or prose" (1988:237). Wordsworth is a romantic idol, whose importance "in popularizing the Lake District, and indeed in synthesizing this transformation of literary place into tourist place, should not be underestimated" (1988:243). His poetry, Squire continues, has

> made each place he mentions a place of pilgrimage, and he has probably added more names than any other writer to a literary map of England. ... Indeed, the images of place that Wordsworth created in his poetry now supersede and transcend environmental actuality (1988:243).

Urry (1995:193–210), in his case study of the Lake District, stresses the important role played by the "place-myth" that English romanticism helped foster around the Lake District. The "development," he writes, "of the Lake District as possessing a particular place-myth only occurred because of visitors and writers and of the incorporation of Romanticism into what has come to be known, taught and revered as English literature" (1995:194). The place-myth, once formulated, is perpetually reproduced by the continual flow of incoming visitors. Indeed the place-myth itself becomes the very attraction of the Lake District.

It should be acknowledged here that the *direct* influence of romanticism upon tourism may perhaps be relatively limited, since not everybody is familiar with or is sympathetic towards the works of the romantics. As Mumford claims, "Romanticism as an alternative to the machine is dead" (1934:287). But the taste for nature that was developed by romanticism is not dead; it has permeated culture as a whole and has often been viewed as a central contemporary value in regard to nature. In this sense, the *general* influence of romanticism is significant and should not be underestimated. To put it another way, romanticism shapes tourism mainly through its general cultural taste, values, preferences, "structures of feelings", and styles in regard to nature. These taste, values, preferences, and so on have been widely adopted by contemporary people. Romanticism as a concrete doctrine is perhaps transcended today, but romanticism as a general taste and regard for nature is never outdated. Rather, romantic taste for nature is spread throughout society as a whole, and constitutes the cultural foundation of contemporary nature, green, or rural tourism.

Romanticism and the Technological Environment

Romanticism has undoubtedly shaped and influenced the development of tourism, both in England and abroad (Buzard 1993:19), but the rise of romanticism in the West (including the romantic taste for scenery) should not be seen as an accidental phenomenon. The romantic movement is an historical phenomenon that, in broad terms, swept across Western and Southern Europe and North America during the 18th and 19th centuries. Its rapid growth implies that the romantic movement has a close relationship with modernity, and technological, industrial, or capitalist civilization. It can be argued that romanticism is a cultural reaction to and a critique of modern capitalist industrial civilization. In Lowy's (1987) terms, Romanticism is "anti-capitalism".

According to Lowy, the central feature of industrial (bourgeois) civilization is "the quantification of life", namely "the total domination of (quantitative) exchange-value, of the cold calculation of price and profit, and of the laws of the market, over the whole social fabric" (1987:892). Surely those are the *societal* aspects of capitalist civilization that romantics such as Rousseau criticized. Discontented with the quantification of life, Rousseau defined nature as an ideal representing innocence and simplicity, expressing the wish to "return to nature", to a simple life. Likewise, what Lowy refers to as "the quantification of life" should also include the impact of the *technological* environment upon organic human life. In other words, the *physical* dimension or the environmental consequence of modernization is also a target which romantics attacked. In a word, romanticism is a cultural demonstration against modern, artificial, urban or technological environments.

Mumford (1934) regards romanticism as a reaction to the civilization of the machine. For him there are two opposing ideas in reaction to this civilization: the romantic idea and the utilitarian idea. The utilitarian favored industrial and commercial civilization. "He believed in science and inventions, in profits and power, in machinery and progress, in money and comfort" (1934:285). Conversely, romanticism was "an *alternative* to the machine" (1934:287). Romantics wanted to restore the vital organic attributes of life to a prominent position because those attributes had been "deliberately eliminated from the concepts of science and from the methods of the earlier technics" (1934:286). Thus, romanticism "provided necessary channels of compensation" for the shortcomings of the civilization of the machine (1934:286) and made efforts towards "the new social synthesis" (1934:287). However, in doing so romanticism often went too far; "it did not distinguish between the forces that were hostile to life and those that served it, but tended to lump them all in the same compartment, and turn its back upon them"; so "it was, for the greater part, a movement of escape" (1934:287).

Mumford classifies the romantic reaction into three groups: the cult of history and nationalism, the cult of nature, and the cult of the primitive (1934:287). All three are in some way related to tourism. The cult of *history* is expressed in the touristic love of relics, ruins, traditions or national heritage (Andrews 1989; Jasen 1991; Ousby 1990). In this sense, romanticism is nostalgic in character (Lowy 1987). The cult of the *primitive* was mostly clearly expressed by Rousseau. Echoed in tourism, it gives rise to ethnic and folklore tourism. Since primitive humans or the "natural" state can not be found in the home society in which modernity dominates, they must be found in remote parts of the world or in "the timeless society". Finally, it is the cult of *nature* that is most relevant to nature tourism. The romantic cult of nature represents an alternative relationship with nature: one that is sensible, affective, and aesthetic rather than utilitarian or instrumental. In Russell's words,

> The romantic movement is characterized, as a whole, by the substitution of aesthetic for utilitarian standards. The earth-worm is useful, but not beautiful; the tiger is beautiful, but not useful. Darwin (who was not a romantic) praised the earth-worm; Blake praised the tiger (1979:653).

Nature tourism, as endorsed by romanticism, is not only a cultural demonstration against but also a cultural compensation for artificial and technological environments. It is not only what Mumford (1934) calls "a movement of escape", it is also a cultural and imaginative creation. If science and technology alienate nature by depriving it of its traditional divine and theological meaning, then touristic romanticism gives nature a new meaning. This new meaning that the romantics bestowed on nature no longer belongs to the divine but to aesthetic sensibilities about nature. Bowie puts this idea well:

> Nothing in the sciences provides a sense of the meaning of nature for the individual subject: the point of science is the production of laws which subsume individual cases. Nature seen with the eyes of modern science starts for many people, though, to look like a machine. Added to this is the awareness that the increasing domination of capitalism leads to nature being regarded in terms of the profit which can be extracted from it. One of the key attributes of the aesthetic is the fact that what makes an object beautiful has nothing to do with its usefulness or its exchange value (1990: 3–4).

Bowie speaks of a notion of modern aesthetics. It can be deduced from this, however, that the meaning of nature is not to be found through a utilitarian perspective on nature (as in the case of science and technology), but through a romantic or aesthetic perspective of nature. It is the romantic orientation that idealizes and re-enchants nature, and hence bestows new meanings on nature: the beautiful, the idyllic, the pastoral, the sublime, and so on.

The romantic re-enchantment of nature should not been regarded as a reaction against specific concrete technological devices or objects, but rather as a reaction against the technological environment *as a whole*. It also does not do justice to romanticism to say that it is hostile to technology *per se*; rather it is fairer to say that what romanticism fights against are the *negative* effects of modern instrumental technology on the physical environment. Technological environments, such as the one represented by the model of the city, are felt to be inhospitable. The culturally based enthusiasm and love of nature and the countryside can only be understood against the background of such technological environments. It is such artificially built environments that account for the romantically idealized and mythologized images of nature and rural scenery. In Williams' (1973) language, the countryside has come to be defined as "pastoral" in terms of the qualities which are absent from urban settings. The idealization of the countryside and the past (based on the new "structure of feeling" exemplified by, among others, British rural-intellectual radicalists) is in fact a cultural reaction to and critique of the civilization of capitalism, industrialism, and urbanism. In other words, without modern technological, urban and industrial environments as a reference, the idealization of the countryside or nature cannot be understood. Mumford makes this point well:

> So long as the country was uppermost, the cult of nature could have no meaning: being a part of life, there was no need to make it a special object

of thought. It was only when the townsman found himself closed in by his methodical urban routine and deprived in his new urban environment of the sight of sky and grass and trees, that the value of the country manifested itself clearly to him (1934:295).

Technology is intended to eliminate scarcity (hence the growth of wealth) and to protect humankind from natural threats and disasters. However, in so doing technology produces another *emergent scarcity*, a scarcity which is characteristic of modern technological and urban environments, namely the scarcity of "natural nature" (Green, 1990:11), of the sights, the ambience, and the amenities of nature. To put it another way, modernity has wiped out the "hunger of stomach". However, it has brought about another kind of hunger, namely, "the hunger of eyes" (Michael 1950b). The modern eye is thus hungry for *natural* nature, which is either beautiful, such as a landscape full of grasses, trees, wild flowers, and wild animals, or sublime, such as vast deserts, huge mountains, and deep gorges. In the urban environment (a product of what Lefebvre (1991) calls "spatial technology"), nature, if there is such a phenomenon, is only an ingredient of urban decoration (parks, gardens, etc.). To a great extent it is driven out of place. It is the scarcity of nature in urban settings that explains the cultural appreciation of nature. "Natural nature" becomes the symbol of a victim of the processes of industrialization, urbanization, and technologization, and therefore becomes the object of pity and nostalgia. Moreover, those parts of nature that avoided the violence of modernity and technology become the new object of worship and pilgrimage. The return to nature is therefore an escape from the hold of the technological and urban environment, or an "escape from the machine" (Mumford 1934:296). In Ousby's words, "as urban life took firmer hold, people would long for nature as a necessary refuge, a source of spiritual renewal" (1990:131). In Dann's terms, if in the Middle Ages the countryside was regarded as a dangerous place, then

> During the overlapping Industrial Revolution, and the subsequent decades which extended well into the twentieth century, it was the city that came to be feared. Moreover, it was against the backdrop of this urban dread that the countryside was projected as a deindustrialized, depoliticized asylum far from the madding crowds, well removed from the toxic waste and pollution associated with urbanization (1997:257).

Nature or rural scenery thus increasingly becomes the sign of simplicity, idyll, authenticity, and amenities, in contrast to the pollution, complexities, and artificiality exhibited in urban and technological environments. Natural nature acts as the "green" dream-place in contrast to the "grey" urban nightmare.

In summary, nature tourism, is not merely the outcome of a general improvement in living standards and of technological advances such as the transport revolution, but also a phenomenon that involves the cultural formation of a taste and preference for landscapes and nature. Once formed, this constitutes a relatively independent causal agent of modern nature tourism. Whereas increased disposable income and time, and technological means of travel are *material* conditions that enable people to seek natural environments through travel based on their taste and preference for nature, taste and preference becomes a *cultural* condition for nature tourism. Without such a cultural condition, nature tourism is not understandable.

Taste and preference must be understood against the background of a technological environment which has its relation to modernity. Romanticism as a cultural phenomenon appeared to be a negative reaction to capitalist, industrialist, commercial, and urbanized civilization, including the artificial technological environment that tends to isolate humans from nature and natural amenities. The separation of humans from nature by technological environments may cause negative psychological and social problems. Thus, there was a romantic reaction to the technological environment by the Romantic movement from the latter half of the 18th century. The origin of nature tourism is closely tied to romanticism and the romantic movement. Although romanticism as a movement may no longer be alive today, its heritage and variants—the taste for and love and appreciation of nature—have been widely adopted by contemporary culture as a whole and are most clearly exemplified by contemporary nature tourism.

Chapter 5

Modernity and Holiday-Making

Tourists are often treated as a homogeneous category. This is misleading; in fact, all tourists are not the same. A most obvious distinction is that between sightseers and vacationers (Cohen 1974). The primary analytical difference between these two types of tourists is "that sightseers seek novelty, while vacationers merely seek change, whether or not this brings novelty in its train" (1974:544–545). Novelty is relative to sightseers' experiences. "A novelty is, in principle, new only once—when one sees or experiences it for the first time" (Cohen 1974:544). Hence, sightseeing tends to be non-recurrent. Change, in contrast, does not necessarily imply novelty. Thus, "one can experience the transition from office work in the city to leisure on the beach as a welcome change, even though it is one's accustomed way of holiday-making" (1974:544). Hence, vacations can be recurrent.

Self-evidently and to a larger extent, the need for change arises within the context of the temporal structure of modernity, since it is this temporal structure which tends to fasten people to a regularized, routinized, and structured everyday life. For those who have full-time jobs, everyday life (including leisure) is usually organized around work and is thus highly temporally structured and routinized. Consequently, work experience is "a key cultural factor" in creating or modifying an individual's need for a holiday (Burns and Holden 1995:41). In this sense, the sociogenesis of holiday-making has to do with the temporal structure of modernity. It is within this temporal structure that the need for change gains significance.

The temporal structure of modernity can also be understood from a phenomenological perspective, that is, from the standpoint of common-sense understanding and experiences. In this sense, time is no longer an abstract concept or form but an amalgam of experiences. Thus, the rhythm of modernity, as exemplified in the case of either Fordism or post-Fordism in the twentieth century, is reflected in people's experience

91

of everyday life. Tourism is a cultural response to the rhythm of modern life. It must be noted that, here, Fordist rhythm and post-Fordist rhythm are understood as two ideal types of work-related experience. Fordism dominated until the mid 1970s, but thereafter post-Fordism tended to gain ascendance. However, they continue to coexist. For example, Fordism still exists in the service industry. In mass tourism, itself a form of service industry, Fordism still survives to a certain extent, though it may be argued that it has lost its dominant position.

Interestingly, when the rhythm of Fordism led to a demand for change and escape through holiday-making, in response to tourists' demands the *organization* of tourism first assumed a Fordist pattern. Its heyday was the age of mass tourism, characterized by the standardization, homogenization, and inflexibility of the product. Although it dates back to the middle of the nineteenth century, mass tourism triumphed in the post-war period. This type of tourism treats tourists as homogeneous. It supplies the tourism package in a standard form and on a large scale. Economy of scale (large scale, lower price) is the goal of mass tourism. Under this the specific and unique demands of individuals are ignored.

However, since the 1980s, various forms of "alternative" tourism have emerged. The tourist market is becoming more segmented and diverse, and demand has become less homogenized. The changes in tourist demand have been identified by Poon (1993 quoted in Burns and Holden 1995:223), as follows:

Table 5.1 Old Tourists and New Tourists

Old tourists	New tourists
Search for the sun	Experience something new
Follow the masses	Want to be in charge
Here today, gone tomorrow	See and enjoy but do not destroy
Show that they have been	Go just for the fun of it
Having	Being
Superiority	Understanding
Like attractions	Like sport and nature
Cautious	Adventurous
Eat in the hotel dining room	Try out local fare
Homogeneous	Hybrid

In response to the new market, producers and suppliers are increasingly supplying tourism in a post-Fordist fashion (Urry 1990a, 1994a). This does not imply that Fordist patterns of mass tourism have completely disappeared. However, as tourists have become maturer and their demands more flexible, diverse, and changeable, the tourist market seems to have become fragmented. So while package beach holidays and other mass tourism products still have a significant market, post-Fordist types of tourism will, it is often argued, be the general trend for the future. Urry has identified the characteristics of post-Fordist tourism as follows:

> consumers are increasingly dominant and producers have to be much more consumer-oriented; the rejection of certain forms of mass tourism (holiday camps and cheaper packaged holidays) and increased diversity of preferences; a greater volatility of consumer preferences with fewer repeat visits and the proliferation of alternative sights and attractions; increased market segmentation with the multiplication of types of holiday and visitor attractions based on lifestyle research; the growth of a consumers' movement with much more information provided about alternative holidays and attractions through the media; the development of many new products each of which has a shorter life so that there is the rapid turnover of tourist sites and experiences because of fashion changes; and increased preferences expressed for non-mass forms of production/consumption such as the increased demand for refreshment and accommodation services which are individually tailored to the consumer (such as country house hotels) (1994a:236).

One of the aims of "alternative" tourism may be to avoid the negative consequences associated with mass tourism. This does not mean that mass tourism is necessarily "bad" and leads invariably and solely to negative consequences. Rather, whether the consequences of tourism are positive or negative is more an issue of planning and management. However, the shift from Fordist ones to post-Fordist ones does reflect, from the tourism supply side, a change in taste and demand on the part of tourists, one which in turn reflects the structural and cultural changes in contemporary Western societies (1994a).

This chapter examines the relationship between holiday-making and the temporal structure, along with the associated experiences of time in modernity. It consists of three sections. The first analyzes the temporal order of modernity. The second outlines the phenomenology of the rhythms of modernity, and discusses how the temporal structure of modernity modifies and creates the need for holiday-making and for a change from everydayness. The third section elucidates how holiday-making is tied to cultural meaning and experienced as escape.

Modernity and its Temporal Structure

Time is an obvious and important factor in the constitution of society, especially in modern society (Giddens 1979, 1981, 1984). But what is time? It is still an enigma (Jaques 1990). Some philosophers (e.g., Descartes, Leibnitz, Kant, Bergson, Heidegger) have defined time in an abstract way, as a form without social content. For sociologists in contrast, time is all about social stories: time is thought of as collective rhythm (Bourdieu 1977; Durkheim 1995; Young and Schuller 1988; Zerubavel 1981), as social time differentiated from natural or astronomical time (Gurvitch 1990; Lewis and Weigart 1990; Mukerjee 1990; Sorokin and Merton 1990), as capitalist time-consciousness (Thrift 1990), as a means of social ordering, regulating, and coordinating (Bourdieu 1977; Lewis and Weigart 1990; Moore 1963; Mumford 1934; Starkey 1988; Zerubavel 1981), as a structuring life-project (Roche 1990; Schutz and Luckmann 1974), as a commodifiable resource and the medium of commodification of both goods and labor (Giddens 1981), as a cultural phenomenon (Bourdieu 1990; Coser and Coser 1990; Kern 1983; Malinowski 1990; Zerubavel 1990), as a symbol referring to the social activity of timing (Elias 1992), and as the complex of various time phenomena (Adam 1990).

In this chapter, a philosophical conception of time will not be adopted. Rather, time will be considered sociologically as a *socially constructed* temporal structure or collective rhythm that establishes order in social life and its activities on four fronts: sequential structure, which tells people in what order social actions take place; duration, which informs persons how long an action lasts; temporal location, which refers to when actions take place; and rate of recurrence, which means how often individuals perform certain actions (Zerubavel 1981:1). This temporal structure must be understood as temporal structuration in a Giddensian sense. It is a collective time-structuring or social timing activity (Roche 1990), but it is simultaneously a temporally structured order. The temporal structure of a society has two interrelated dimensions. First, it involves an institution or organization of time, which is exhibited as collective rhythm, schedules, or pace of life. Second, it involves a *"time habit"*. While this time habit is shaped by a given temporal structure in society, it produces or reproduces that structure at every moment. A time habit can be either reflexive or non-reflexive. At the reflexive level, time-habit is time conception or time consciousness.

Industrialization and the Modern Rhythm

Generally speaking, modernity has dramatically transformed social time. In traditional societies "the experience of time is not separated from the substance of social activities" (Giddens 1981:9), whereas in modern societies, clock time exists alongside experienced time. This "public, objectified time of the clock" functions as the "the organising measure of activities of day-to-day life" (1981:9). The emergence of clock time (abstract, objectified time) indicates the increased complexity of society (based on the division of labor), which entails temporal coordination and synchronization. In his essay "The Metropolis and Mental Life", Simmel (1950b) analyzes how the appearance of impersonal time plays an important role in the structuring of urban life and consequent alienation. Indeed, modernity would not come into being without a corresponding modern time or modern temporal structure. This temporal structure exhibits four major characteristics.

Synchronization. The advent of industrial civilization implies an increase in organic solidarity, system complexity, division of labor, and the urgency of temporal synchronization and coordination within and between various organizations and their functional parts (Elias 1992; Hassard 1990; Moore 1963; Starkey 1988; Thrift 1990; Zerubavel 1981). Thus, Mumford argues that it is the clock, not the steam engine, that is the key machine of the industrial age (Mumford 1934:14). Time is thus a medium of synchronization and coordination in the modern industrial age. The synchronization of social activities becomes more self-evident under the condition of late modernity. Moreover, such a synchronization has been extended to the global level due to advances in the technology of telecommunication (Friedland and Boden 1994) and an increasing "separation of time from space" (Giddens 1990). The advent of post-Fordism assigns to time a much more crucial role. Flexible accumulation, ephemeral consumer demand, reductions in turnover time in production, the increasing significance of "just-in-time" inventory systems (cf., Harvey 1990), all presuppose time as a crucial medium in concerted actions and in the production and reproduction of social order.

Pace of Life. On the one hand, the modern industrial system employs machines and modern technologies in the production process. Without humans the machine is dead. The birth of modern industry required a new time-habit, if machines and humans were to be integrated (Thrift 1990:114). The factory imposed an artificial "time discipline" upon humans (Marx 1954; Thompson 1967:90), which then brought about a new pace of life. Humans were, to some extent, forced to follow the rhythm of the machine rather than vice versa (e.g., on assembly lines). On the other hand, the modern pace of life is an indicator of the increased complexity and flexibility of society as a whole.

Efficiency and Productivity. Under the conditions of industrial capitalism, the drive for profit requires that work be scientifically designed and managed in terms of time and the efficiency principle (Hassard 1990). Taylorism, Fordism, and the more flexible post-Fordism all regard efficiency and productivity as their principal goals.

Routinization. As mentioned above, time is a very important medium of synchronization (based on the division of labor), pace of life (based on machine production and a money economy), and efficiency (based on scientific management). Time under modernity becomes an "ordering principle" (Zerubavel 1981). Its result is the temporal structure of day-today life, which appears as routine. On the one hand, work, at least a considerable amount of work, has been routinized, as illustrated, for example, by work on assembly lines, in bureaucratic offices, in super-markets, in hotels, and in fast-food chains. On the other hand, everyday life as whole has also been routinized and organized around work, daily commuting and housework. Thus, there arises a routinized separation between working time and free time.

Routinization, moreover, can be considered a universal phenomenon. It exists not only in modern society but also in traditional society, albeit with certain differences. In traditional agricultural society, daily routines were regulated by *natural rhythms* or seasonal tasks rather than by an artificial timetable (Bourdieu 1990; Mukerjee 1990). There was no clear institutional *temporal* demarcation that separated work from leisure. In the busiest season, such as harvest time, the length of the working day was extended to such an extent that the time left for sleep was far from adequate, whereas in a slack season, such as winter, the whole period could be used as leisure time. Moreover, once routines were stabilized as tradition, little could be allowed to change, for tradition had become a moral authority to which later generations were subject. By contrast, within modern society, as Giddens argues, routinization in modern life is largely subject to the dictates of "dull economic compulsion" (1981:11). It is economic rationality, rather than natural rhythm, that makes time function as a segmenting principle, that separates one kind of activity from others, for example the private sphere from the public sphere, and work from leisure (Hassard 1990:7; Moore 1963; Zerubavel 1981:141). The oscillation between work and leisure—or the routinization of daily life—under modernity is thus organized around the rational way of allocating "objectified" time units which the modern organization of capitalist production entails.

Rationality, which is embedded in the modern industrial order, there-fore brings about a fundamental transformation of social timing, the temporal order and the structure of everyday life. As a result, modern routinization embodies a relatively stable form of temporal order and structure. It is through routinization that social structure is constantly reproduced. Time is thus an integral element of the social construction of reality. In brief, modernity, in the name of efficiency and instrumental rationality, brings about a modern rhythm that replaces traditional rhythms. This modern rhythm is enabling because it gives rise to greater efficiency, productivity, and constant innovation. It is also constraining because it brings about deadening routines and stressful deadlines. Modern rhythm is therefore ambivalent; it gives people order, but at the expense of spontaneity (Zerubavel 1981:47).

The Commodification of Labor and Modern Time Consciousness

Modernity brings about not only a new temporal order (schedules, routinization, pace of life and collective rhythm, etc.), but also a new consciousness of time (Thompson 1967; Thrift 1990). Quite different from the people of traditional agricultural societies, who assumed a *task-oriented* time habit, modern people share a common conception of time. Time is no longer confused with substantial experiences but rather regarded as linear, quantitative, objectified, measurable, and divisible time—as abstract clock time. Most importantly, people in the context of modernity widely adopt the dictum that "time is money", an idea that Weber quoted from Benjamin Franklin in order to exemplify the "Protestant ethic" (Weber 1970:48). As Lakoff and Johnson (1980) suggest, the modern conception of time is illustrated by three widely accepted metaphors—time as money, time as a limited resource, and time as a valuable commodity.

Such conceptions of time have obvious connections with the *commodification of time* which characterizes modern capitalism. According to Giddens (1981), capitalist commodity production is made possible by the prevalence of two processes of commodification: the commodification of goods and of labor. Goods and labor power become interchangeable commodities. What permits this interchangeability is the commodification of time itself, which acts as an underlying medium of these two processes. "'Commodities' exist only as exchange values which in turn presuppose the temporal equation of units of labor" (1981:8).

Time, as perceived by modern people, has thus gained emergent values, properties, and features. As Hassard observes:

> one is exchanging time rather than skill: selling labor-time rather than labor. Time becomes a commodity to be earned, spent or saved.... Time had a value that could be translated into economic terms.... Time was a major symbol for the production of economic wealth (1990:13).

Historically, such a conception of time was first learnt by capitalist entrepreneurs and then gradually taken over by the working class, as a weapon in the labor movement. Relatedly, Thompson notes that:

> The first generation of factory workers were taught by their masters the importance of time; the second generation formed their short-time committees in the ten-hour movement; the third generation struck for over-time and time-and-a-half. They had accepted the categories of their employers and had learned to fight back with them. They had learned their lesson, that time is money, only too well (1967:86).

Time, then, becomes both a medium of increasing organic solidarity (e.g., synchronization) and a medium of weakening social integration in Lockwood's (1964) sense (e.g., chronic class conflicts). If *industrializa-*

tion largely gives rise to increased temporal synchronization and organic solidarity (hence shapes people's time-habits), then the *commodification of time*, underlying the commodification of both labor and goods, plays a considerable role in changing people's time consciousness. Time becomes a resource for which both the upper class and the working class struggle.

For employers, workers are the commodities of labor power; they are bought from the labor market, but only for limited use in the twenty-four hours of each day (it is a banality that there is an absolute limitation in labor time because of biological needs—sleep, eating, etc.). Employers use these purchased commodities to the maximum possible level. Thus, they have to calculate carefully the most efficient way in which their workers should use their working time. Therefore, theoretically speaking, in the eyes of the employers, employees are only the means to an end—the maximization of profit. They must be used properly and efficiently. Since the exploitation of workers is realized through obtaining the surplus labor time that creates surplus value or so-called "profit" (Marx 1954), there are two ways of squeezing out surplus value. As Marx argues, under early capitalism "absolute surplus value" was the major source of profit and was gained by maximizing the absolute length of the working day. With an increase in productivity, and partly due to the pressure of the labor movement, the source of profit then mainly came from the "relative surplus value" that was created through intensifying the pace of work and increasing the efficiency and productivity of labor (see Starkey 1988). As for leisure, from the standpoint of capitalists, employers, or managers it is needed only on the ground that it allows workers to recover from physical and mental fatigue and thereby improve their work efficiency and productivity. For them, leisure time has to be controlled; it should not be used to engage in getting drunk or any other hedonistic activity.

For employees the story is altogether different. Work is only a means of earning subsistence, denuded of inherent meaning. In the typical words of one laborer, "The things I like best about my job are quitting time, pay day, days off, and vacations" (Chinoy 1955:85). Working time, then, for most employees, especially workers on assembly lines, is alienating time. It is in free time that they feel at home (Marx 1977). Thus, it is no wonder that the length of the working day or the number of weeks worked per year also becomes a target which the working class use their labor movements to try to reduce. Leisure time becomes one of the goals for which workers struggle. If, under early modernity, leisure was the upper class's privilege (Veblen 1925), then based on increases in productivity and with the unfolding of the class struggle, universal leisure has gradually become a higher priority (Pimlott 1976).

Time is money and value. This is an emerging modern consciousness. Time thus becomes one of the essential elements of class bargaining, conflict, and struggle, of industrial relations in the labor movement (Starkey 1988:101). Capitalist employers wish to extend the length of the working day and increase the number of weeks worked. Employees, however, wish to reduce these. Thus, generally speaking, the actual length of the working day or the number of weeks spent in working is in fact the result of a power balance or functional interdependence (to borrow Elias' term) between both sides, as well as regulation by the state. Historically, the general tendency is towards a constant increase in leisure time.

The Emergence of the Institution of the Paid Holiday in Britain

One of the essential elements of the labor movement and industrial relations that directly relates to tourism is paid holidays. Holiday-making, looked at from a historical perspective, is not merely the result of increased productivity and enhanced living standards, but also the product of collective bargaining and class struggle. As Böröcz observes:

> A primary focus of working class struggles in the mid-nineteenth to early-twentieth centuries was precisely the issue of the reduction and regulation of labor time, that is, rephrased from our point of view, struggle for the provision of ample free time to be expended on leisure activities. The standardization, normalization, and commercialization of free time is one of the most obvious outcomes of this struggle. Thus, industrial capitalism is a key factor in the emergence of the institution of leisure migration (1996:28).

Pimlott (1976) has traced the social history of the Englishman's holiday. In England, domestic holiday-making can be traced to the 17th century visits to spas (e.g., Bath) and seaside resorts (Blackpool, Scarborough, etc.) for the purpose of improving health (Hern 1967; Urry 1990a: 16–39). International tourism can be traced to the same century's Grand Tour. These journeys were the privilege of only the wealthy elite at that time, and the majority of the masses could not afford them. Only with the development of the railways, increasing income, and the emergence of the economic organization of travel by train at a low price, as shown in Thomas Cook's tourism enterprises, could a greater number of people afford to participate in tourism and seaside holidays. Holiday with pay, however, was not merely the product of technological advance (the rail age), an increase of income, or the emergence of commercial organization (e.g., the tours organised by Thomas Cook), though these factors were important; it involved an extra factor—industrial relations and class struggle. Holiday with pay began in the second half of the 19th century, propelled particularly by The Bank Holiday Act of 1871. Although in the 1880s holiday with pay as a practice began slowly to spread, it was mainly the privilege of the middle classes rather than manual workers because it involved growth in the real wages of workers, something which employers were reluctant to offer (Pimlott 1976:214). It was not until the 1930s that paid holidays spread to a wider range of classes and became more of an institutionalized practice. The major obstacles to its diffusion were the capitalist employers' opposition and the government's reluctance, but pressure from below (the working class and trade unions) and from without (international pressures, e.g., statutory provisions for holiday with pay in foreign countries) gradually forced employers and politicians to accede to the demands of workers.

In a report addressing the issue of whether holiday with pay should become statutory, presented by the Ministry of Labour to Parliament in 1938, it was suggested that "an annual holiday contributes in a considerable measure to workpeople's happiness, health and efficiency" (Report 1938:54). The report treated "health" and "happiness" as contributory factors to industrial efficiency, and claimed that the current increase in industrial productivity was so great as to be able to absorb in a short time the cost of holiday pay, arguing that the full benefit of increased productivity should go to the workers (1938:25). The report also pointed out that for various reasons there was opposition by employers and, the National Confederation of Employers' Organizations to any statutory enactment of holiday with pay, despite the fact that it already existed to a limited extent. The main reasons for this opposition were that it would increase the burdens of industry and that its cost would adversely affect the competitiveness of British industry *vis-à-vis* foreign industries (1938:28–34). However, the report strongly recommended "that an annual holiday with pay should be established, without undue delay, as part of the terms of the contract of employment of all employees" (1938:60). The holiday, it was suggested, should consist of at least as many days as are in the working week and these days should be taken consecutively (1938:60).

The sociogenesis of holidays with pay mirrored the emergence of socially, culturally, and politically accepted values concerning health and happiness, which were seen as leading to industrial efficiency and inproved industrial relations. Relatedly, there also emerged a modern leisure consciousness, that is, there was "a change of mental attitude" with regard to leisure (Pimlott 1976:23). Leisure travel or holiday-making was no longer seen as a waste of time, and contradictory to the work ethic. Rather, it was viewed as a necessary complement to and a compensation for work, as a means of enhancing productivity and efficiency, and as an essential element of a reasonable standard of living (Pimlott 1976). If the commodification of time led to temporal alienation, then, paid holidays were an institutional antidote to and compensation for such alienation.

The Phenomenology of the Modern Rhythms

Phenomenology, when applied to sociological studies, is a perspective where social life is examined from the standpoint of everyday common-sense understanding and "native" social actors. Time can be investigated not only from a structural perspective as above, but also from a phenomenological perspective (see Roche 1973, 1990; Schutz and Luckmann 1977), in order better to appreciate how people actually experience the time, tempos, and rhythms of modernity. Two points need to be noted here. First, although some phenomenological sociologists use terms that lay persons never employ, these expressions are nevertheless valid to describe lay persons' experiences since they involve "second order" interpretations of their "first order" experiences (Giddens 1976). Second, on an experiential level, time ceases to be a purely abstract concept; rather "clock time" is reified and forms part of the experience of social processes. Hence, a persons' phenomenological experience of time may contain "substances" of other experiences (at the prereflective level). Time is not only conceived of as *empirical* clock time, timetables, schedules, deadlines, routines, time budgets, and punctuality, but is also associated with various "feelings" that involve the "quality of time", such as "busy" or "free", a "good" time or a "bad" time. Monotony or novelty, boredom or excitement, slowness or speed, ease or urgency, certainty or uncertainty, security or anxiety, all involve a "feeling" of time, tempos and rhythms.

The modern pace of life has shaped people's feelings of time. For those enduring long-term unemployment, time is a burden (Roche 1990). For those who are a "cog" in the economic machine, time is a source of pressure. Thus, modernity includes the phenomenological experience of modern time and rhythm. As Roche explains, modern people may "feel that they 'have' either too much or too little time, situations in which the *pace* of life seems either too fast or too slow for comfort" (1990:73). In other words, modern rhythm and tempo often cause psychological discomfort. Time is a typical source of ambivalence within modernity.

The modern pace of life reflects the structural determination of modernity. Collective rhythms, rigid schedules and timetables, stressful deadlines, and fast tempos all constitute the structuring temporal order to which members of society are subject. Moreover, personal socialization entails temporal socialization which makes individuals adapt to a given collective rhythm and failure to keep up with this results in career failure. Furthermore, the modern temporal order is reinforced by the collective values and norms in modern societies. For example, the Western value of individualism, the capitalist ideology of free and fair competition, and the Protestant work ethic are all in one way or another motivational forces whereby individuals aim to keep up with the normal tempo, or at least not to fall behind it. In Bourdieu's terms, they must have "respect for collective rhythms" (1977:162).

The Phenomenology of Routinization

Recently, a debate has arisen on the question of whether the West has undergone or is undergoing a transition from "Fordism" to "post-Fordism". This controversy is related to the debate on the broader issues of modernity and postmodernity. Usually, Fordism is linked to modernity and post-Fordism to postmodernity. The debate is ongoing, but it is believed that Fordism has not yet entirely disappeared; many of its elements are still alive and coexist with post-Fordism (such as McDonaldization) (Ritzer 1996:150–153). Both Fordism and post-Fordism can be treated as analytical devices. Each has its own rhythms. Ideal-typically speaking, the work rhythm of Fordism can be characterized by routinization. By contrast, the work rhythm of post-Fordism is typically exhibited by the uncertainty that is associated with flexibility. Even so it is not denied that Fordism can also display uncertainty, just as post-Fordism can also contain an element of routine.

Fordism—named after Henry Ford, inventor of the assembly line, which is representative of the modern mass-production system—grew throughout the 20th century in the West, reaching its peak in the 1950s and 1960s, and showing signs of decline from the mid 1970s and during the oil crisis of 1973. The major features of Fordism include the mass production of homogeneous products, the use of inflexible technologies (such as the assembly line), economies of scale, the deskilling, intensification, and homogenization of labor, a mass market for the homogenized products of mass-production industries (Ritzer 1992:313; Kumar 1995: ch.3). Fordism also appears to be characterized by the principle of full and long-term employment with high wages (Keynesian economic policies). Clearly the period in which Fordism triumphed was also the time in which mass tourism reached its peak. There is a close link between Fordism and mass tourism. On the one hand, mass tourism is itself a Fordist pattern of tourism (the homogenization of tourist experiences). However, mass tourism emerges as a cultural response to working conditions and the rhythm of Fordism (alienation, monotony, routinization, deskilling, etc.). The phenomenological experience of time and tempo in Fordism is certainly one of the reasons for the emergence of the democratic demand for "escape" and for "periodical change from the routine and environment of everyday life" (Pimlott 1976:213).

Alienation due to the Routinization of Work. In the age of Fordism the experiences of time and work by the working class, particularly those long-term employees who toiled on the assembly line, were ambivalent. Individuals needed to work because they had to earn a living. In this sense work had a meaning for a laborer. However, the experiences of work and the rhythms associated with work could also be negative. Time relating to work was for the employee perhaps too rigid, too structured, too routinized, or too monotonous. Fragmentation, monotony, and alienation were the main problems resulting from the routinization of work.

There is abundant literature on this topic. If Durkheim's *The Social Division of Labour* (1964) implied the notion of the routinization of work, then Max Weber's (1978) theory of bureaucracy and the "iron cage" was a clear and direct example of its existence (i.e., the bureaucratization of work). Another classic example of such studies is Adam Smith's (1910) discussion of the methods of reducing costs and increasing output in a pin factory through the detailed technical division of tasks. In later industrial practices, informed by the doctrines and principles of Taylorism and Fordism, the routinization of work became part of a compelling and fundamental managerial logic.

Most studies of the topic centre on manufacturing work (Beynon 1973; Blauner 1964; Braverman 1974; Burawoy 1979; Walker and Guest 1952) and clerical work (Braverman 1974; Garson 1975; Mills 1951), but more recently the subject matter has been extended to include "interactive service work", such as work in McDonalds or Combined Insurance (Leidner 1993; Ritzer 1996). As suggested by these studies, the routinization of work has become widespread since it brings many benefits to both capitalists and managers—increasing efficiency, cost reduction, uniform quality of products, and increasing control over workers and the working enterprise (Leidner 1993).

Several characteristics of this routinization have been identified. First, work is broken work down into routines, specifically, the fragmentation of tasks is based on the division of labor, which confines workers or clerical staff to narrowly defined aspects of production or work. Thus, work is transformed into a simplistic, monotonous, and repetitive process. This process is most vividly described by Gillespie:

> In industry the person becomes an economic atom that dances to the tune of atomistic management. Your place is just here, you will sit in this fashion, your arms will move x inches in a course of y radius and the time of movement will be .ooo minutes (Gillespie 1948; quoted in Fromm 1956:125).

Second, in association with routinized work, workers or clerical staff are deskilled, degraded, and devalued. Gillespie continues:

Work is becoming more repetitive and thoughtless as the planners, the micromotionists, and the scientific managers further strip the worker of his right to think and move freely. Life is being denied; needs to control, creativeness, curiosity, and independent thought are being baulked, and the result, the inevitable result, is fight or flight on the part of the worker, apathy or destructiveness, psychic regression (Gillespie 1948; quoted in Fromm 1956:125).

Third, workers and clerical staff are denied control over the overall production process or overall work function. In Braverman's (1974) terms, one of the principles of the routinization of work is the separation of conception from execution. Workers or clerical staff are excluded from initiative, creativity, and responsibility for conceiving, planning, and designing their work tasks. This, as Simmel points out in *Philosophy of Money* (1990) makes employees feel that work is external to and estranged from their own lives. "When work is not inherently involving, it will be felt as monotonous" (Blauner 1964:28). Thus, such work increases the possibility of subjective monotony.

Fourth, workers and clerical staff face constraints in quitting their jobs. Some people accept work not because it is interesting, but simply because they have no other choice. Employees may have to, or are socialized to, accept and tolerate fragmentation, monotony, and estrangement from work for the sake of material rewards (wages) (Goldthorpe and Lockwood 1968). Discursive satisfaction with a job does not mean that there is no psychological discomfort in work; rather, such discomfort is counted as a necessary sacrifice in order to earn a wage. However, such constraints on available work may trigger off demands for complementary or opposite experiences, usually free, individuated, interesting, and exciting experiences, in the world of leisure.

All these features are the embodiment of "alienation" in the routinization of work (Blauner 1964). According to Blauner, alienation is a "quality of personal experience which results from specific kinds of social arrangements" (1964:15) (for a conceptual clarification of the category of alienation see Ludz 1973). Such alienation has its structural sources, such as the routinized pace and rhythm of industrial work. Blauner identifies alienation in four areas. First, there is powerlessness. This is the situation in which a person is downgraded as "an object controlled and manipulated by other persons or by an impersonal system (such as technology)" (Blauner 1964:16). Second, there is meaninglessness, the situation where a worker may "lack understanding of the co-ordinated activity and a sense of purpose in his work" (1964:22). Third, there is isolation, that is to say, the workers lack a sense of membership of an industrial community or organization. Fourth, there is self-estrangement, the notion "that a worker may become alienated from his inner self in the activity of work", and the "consequence of self-estranged work may be boredom and monotony" (1964:26).

All four of these dimensions involve the routinization of work (repetition, monotony, deskilling, etc.) in industrial environments. Routinized work in industry or a bureaucracy imposes a constraining, compelling, and rigid tempo and rhythm, a situation in which individuals become automated, robot-like, de-individualized, repetitively doing Sisyphus-like wearing tasks. Toiling under such a working rhythm, employees' acts, pace, and speed are set by machines and managers' scientific calculations, and must be geared to the requirements of machines or scientific management (Mukerjee 1990). Under such conditions workers experience temporal alienation as well as structural alienation.

Reflections on the Routinization of Life. People not only experience the malaise caused by the routinization of work. They may also pause and reflect on the meaning of the routinization of life as a whole that is dictated by work. Certainly, routine is not itself "bad". On the contrary, it is fundamental to life. As Anthony Giddens argues, "routine occupies a very important place in the reproduction of practices" (1979:218). Routinization gives rise to "ontological security", taken-for-granted order and a psychological sense of safety (Giddens 1979, 1990, 1991). Ontological security "is sustained primarily through routine itself ... when routines, for whatever reason, become radically disrupted ... existential crises are likely to occur" (Giddens 1991:167). Clearly, without such an ontological security, life would be full of anxiety and fear.

However, routine is also ambivalent. Although it leads to a sense of security, it can also bring about a sense of boredom and monotony, as demonstrated by Cohen and Taylor (1992) (for psychological studies of boredom see Apter 1992; Berlyne 1960; Csikszentimihalyi 1975; Fenichel 1951; Geiwitz 1966; Hill and Perkins 1985; Mikulas and Vodanovich 1993; O'Hanlon 1981; Perkins and Hill 1985). "Why is each day's journey marked by feelings of boredom, habit, routine?" ask Cohen and Taylor (1992:46). For them, this situation relates to the routine of life, where people are "forced" to exist in conformity with "paramount reality". Routinization leads to "the predictability of the journey", to the knowledge "that today's route will be much like yesterday's", and therefore to "an awful sense of monotony" (1992:46). According to Cohen and Taylor,

> the ideal state is some equilibrium between the routinized and indeterminate aspects of our existence; if too many of our actions feel repetitive and determined then we become "fed up" and feel the need to break away a little, to shake off some routines (1992:49).

However, modernity often overdetermines its pace of life, and work often obliges individuals to structure their daily lives into routines.

Therefore, people may sometimes be obsessed by a sense that is so ubiquitously routine that not only everyday life, but existence itself, is trapped in the prison of the mundane, the trivial, and the readily predictable. "This is the experience we call boredom, monotony, tedium, despair" (1992:50). These feelings may be corrosive. They threaten individuality and identity. Thus, people may feel that "something must be done about it" (1992:48). Rather than seeking a solution in paramount reality itself, they "look elsewhere to cope with routine, boredom, lack of individuality, frustration" (1992:112). Individuals search for "free areas", somewhere to flee to and temporarily escape paramount reality, anticipating "that its arrangements must be toyed with, its rules relaxed, its gates opened" (1992:113). Cohen and Taylor classify three different free areas: activity enclaves, which include hobbies, games, gambling, and sex; new landscapes, which consist of holidays and adventures; and mindscape, which refers to drugs and therapy. All these free areas convey the escape message, the message of the will to escape the mundane, the routine, boredom, and paramount reality.

Cohen and Taylor demonstrate the possibilities of encapsulating tourism and holidays in terms of the sociology of everyday life. Holidays are "temporal excursions away from the domain of paramount reality" (1992:131). Such excursions are best understood in terms of the effects of paramount reality, including boring routines. This approach of boredom-avoidance can also be found in many other researchers' studies, for example in Krippendorf's (1987:ch.10) study of the holiday-maker, in Mitchell's (1983, 1988) exploration of mountain experiences, and in Elias and Dunning's (1986) analysis of leisure and sport.

Varying from Cohen and Taylor, Camus (1955), in a reflection on the routinized rhythm of life, treats constraining routines not as boredom but as "absurdity". The absurd is a sense of the meaninglessness of life, which is derived from reflection on the meaning of life. For Camus, such reflection often leads to a questioning of routines that are taken for granted as natural and normal and thus reveals the "ridiculous character" of routine life, namely "the absence of any profound reason for living" (1955:13). Camus writes:

> It happens that the stage-sets collapse. Rising, tram, four hours in the office or the factory, meal, tram, four hours of work, meal, sleep and Monday, Tuesday, Wednesday, Thursday, Friday and Saturday, according to the same rhythm—this path is easily followed most of the time. But one day the "why" arises and everything begins in that weariness tinged with amazement. "Begins"—this is important. Weariness comes at the end of the arts of a mechanical life, but at the same time it inaugurates the impulse of consciousness. It awakens consciousness and provokes what follows. What follows is the gradual return into the chain or it is the definitive awakening. At the end of the awakening comes, in time, the consequence: suicide or recovery (1955:18).

Therefore, a sense of the absurd occurs when the "mechanical" routine life is called into question. Here Camus reveals a *reflective absurdity*, an absurdity that is derived from the collapse of meaning and purpose in mechanical routines. Life under modernity can become a problem. Therefore, against this background, it is understandable that tourism or a holiday, like religion and pilgrimage, functions as an institution that bestows meaning on life (Graburn 1983a).

The Phenomenology of Uncertainty

From the 1970s onwards post-Fordism gradually emerged as a new order of the political economy. The features associated with post-Fordism include: segmentation of the market with the growth of interest in more specialized products, consumer-led, more flexible specialization in production and shorter production runs, taking full advantage of new technologies in order to make profits, a destandardization of labor which entails additional diverse skills, more responsibility, and greater autonomy, and hence leads to the differentiation and individualization of the labor force (Kumar 1995:36–65; Ritzer 1992:314, 1996:151–152). Furthermore, if Fordism adopts a *long-term* employment strategy, such a strategy can increasingly be replaced by a pattern of short-term employment contracts and part-time work, thereby increasing the risk of unemployment (Beck 1992).

The uncertainty, flexibilty, accelerating tempos and rhythms, and the growth of social risk (e.g., unemployment) linked to post-Fordism lead to a different kind of experience of time. If in the Fordist age the major problem for workers was alienation, monotony, and boredom, then in the post-Fordist age, flexible, uncertain, and destandardized work can be extremely challenging. For many the challenges of work can lead to self-fulfilment. However, the other side of the coin is not pleasant: stress, pressure, and anxiety may overwhelm people. Indeed, such experiences of work and time may lead to several psychological and social problems. True, these experiences are subjective and psychological. However, a collective concern for them is a social fact, indicating the seriousness of the problem. According to an electronic literature search (by means of BIDS ISI—Bath Information and Data Services, the social sciences section, UK) for the period 1981 to August 1996, in the journals included by BIDS there are nearly eleven thousand article titles containing the key word "stress", over two thousand including the key word "pressure", and nearly seven thousand containing the key word "anxiety". These statistics certainly do not exhaust all the academic and public treatments of "stress", "pressure", and "anxiety" (for example they do not include book chapters), but these figures are nevertheless quite telling. They indicate that stress, pressure, and anxiety are really serious endemic problems in modern life (Cooper and Marshall 1980; Cooper and Payne 1978; 1980; Forgays, Sonowiski and Wrzesniewski 1992; Goldberger and Breznitz 1993; Gray 1971; Levitt 1968; Spielberger 1972).

Stress, pressure and anxiety are also related to modern "time panic". According to Lyman and Scott, time panic

> is produced when an individual or a group senses it is coming to the end of a track without having completed the activities or having gained the benefits associated with it, or when a routinized spatio-temporal activity set is abruptly brought to imminent closure before it is normally scheduled to end (1970:207).

One of the causes of "time panic" is a series of scheduled deadlines and timetables. In contemporary society, especially under post-Fordism, temporal requirements are more constraining than ever before, since post-Fordism entails not only synchronization but also "just-in-time" production and delivery. As a result, individuals or groups face increasing stress and pressure. Time panic is one of the major discomforts of contemporary life. It is therefore evident and understandable that people suffering from chronic stress and worry need change and relaxation via holiday-making.

For many individuals, including workers, stress grows under the condition of post-Fordism. For managers, executives, and professionals, stress and pressure may be even higher. For them the tempo of life is, or often has to be, very fast because of their higher organizational responsibilities. As Wachtel (1983) observes, over some fifty years, as the number of hours worked has consistently decreased for the labor force at large, the number of hours worked by managers and professionals has actually increased. "In many western countries, professionals and managers now work close to 50% more hours a week than ordinary workers" (1983:46). Why do they work more hours when they are in a position to set their own hours of work? Part of the reason is that "they enjoy their work more. Much of the work of executives and professionals is challenging and stimulating, and leaves room for creativity and self-expression" (1983:46). But for Wachtel this is far from the whole story. A more important reason behind such pressure is the result of competition and the high value placed on individualism. As Wachtel states:

> we live in a highly competitive and individualistic society, and the pressures on us to strive, to achieve, and "get ahead" are enormous, moreover, when everyone else is racing to get ahead, not to do so is to fall behind. Although the advantages of being able to set one's own working hours, to determine when and how much one will work, are obvious, there is a compensating price to be paid as well—having continuously to face the question "Am I doing enough?" and, for many, *never* quite having the sense of one's work being done and it being time to relax (1983: 47).

Thus, even if enjoyment of work is real, "the pressured quality often remains and reveals itself in psychological and psychical symptoms and in a generally high level of tension and irritability" (1983:47). The fast tempo and the stress of work are therefore mutually reinforcing. Stress (responsibilities, competition, success, reputation, etc.) enforces a faster tempo (doing more hours of work, working harder, etc.), and increased tempo in turn brings about further stress. Thus, the only way of escape is to get out of the rhythm. Therefore, holiday-making becomes a functional institutional respite from the stressful tempo of life. It is no wonder that former American President George Bush continued his preplanned holiday when the the Gulf War broke out in 1992.

Holidaying and the Differentiation of Time

Holiday-making involves a particular way of spending free time (Böröcz 1996:28; Ryan 1997b; Vukonić 1996:14). It "presupposes that free time be regulated and packaged in weekly and annual blocks" (Böröcz 1996:28). Thus, the sociology of holiday-making can in a sense be translated into a sociology of free time. Although the concrete forms of holiday-making are various they all share a common feature—they all "consume" a concentrated block of free time annually or semi-annually.

Changes in the Social Ethic of Free Time

Greater efficiency indicates a higher output within a given unit of time (temporal input), namely enhanced productivity. Many factors contribute to the growth of productivity and efficiency of work under modernity, such as the introduction of new technologies, the division of labor, and scientific management. Among other factors, the rational use of time (time management) is also significant. Under modernity time is seen as a resource, commodity, and value, and hence is subject to the principle of rationalization. The temporal order of modernity is rational in the sense that it goes beyond natural rhythms, and is mainly subject to the dictates of economic constraint.

An increase in personal leisure time has certainly to do with higher efficiency and productivity as well as industrial relations. Self-evidently, once a society can produce many times more output for a given temporal input than before, this situation indicates that the production of a given amount of output requires much less time (temporal input) than previously. This increased productivity implies that there exists a potential for increased leisure time. As mentioned above, unionization, together with other factors such as the intervention of the state, has turned this potential into reality. Thus, the post-war era has witnessed the emergence of "universal leisure" in many economically advanced nations. "Paid leave, holidays, and free time are actually written as rights into the constitutions of 65 nations, while others reflect rights to leisure and travel in their legislation" (Richter 1989:16). It is claimed that the West has entered the so-called "leisure society". The emergence of holidays with pay, then, is an embodiment of "universal leisure".

The generalization of holiday-making implies an overall growth of disposable income and free time, on the one hand, and a change in the social ethic of free time on the other. As Pimlott puts it, one of the essential requirements for the rise of modern holiday-making is

> that there should be an increase in the number of persons with a surplus over bare subsistence which they could spend on amenities; and that they should want to spend part of any such surplus on holidays (1976:212).

Thus, the emergence of mass holiday-making entails not only the "declining marginal urgency" of survival caused by affluence (Galbraith 1958) but also a change of mentality and ethic regarding time and income, that is, people want to spend part of their increased free time and increased income on holidays. Seen from the perspective of the work ethic, holidaying is non-productive and hence waste of time. However, with the increase in disposable time and income the negative side of the temporal order under modernity is gradually defined as "intolerable", and hence holiday-making is no longer regarded as a waste of time. It is instead seen as necessary to personal welfare, informed by a changing ethic regarding the non-productive use of time.

Therefore, the rise of institutional holiday-making implies a "normative differentiation of time", that is, a relative short period of time *set aside* from the mainstream tempo and rhythm for a non-productive purpose that is accepted by society and culture. The differentiation of time is related to different ethics. If working time is informed by a work ethic (doing things), then leisure or holiday time is dictated by the need for relaxation, change, play, and freedom. Holiday time is culturally, socially, and politically approved of as exempt from work, obligations, and the work ethic, and may be used for non-productive purposes. Holidays away from the home for the purpose of pleasure are no longer regarded as contradictory to the work ethic, but rather as complementary to it (Pimlott 1976).

A change in the social ethic of free time is the outcome of modernization, including its temporal form. It is the negative experience of the temporal form, as well as other aspects of modernity that produce the desire for differential time, namely holiday time. As Pimlott writes:

> Many forces contributed to put holidays high amongst the items on which increased incomes were spent. The faster tempo, higher degree of mechan-

ization, and enhanced scale of modern industry increased the desire for periodical change from the routine and environment of everyday life (1976:213).

It is against this background of modernity, particularly its temporal order, that a change in attitudes regarding the non-productive use of time can be understood. In addition, modernity also creates sufficient material conditions to sustain a short concentrated period of time each year for holiday purposes.

Holiday-Making as an Institution of Escape

Holiday-making is not only an escape from the space of the home society, but also an escape from the tempos and rhythms of home. Thus, it has its own temporal boundaries. As Graburn observes, a holiday

> is limited in duration and is a contrast with the longer periods of ordinary life. Thus it has a beginning, a period of separation characterized by "travel away from home;" a middle period of limited duration, to experience a "change" in the non-ordinary place; and an end, a return to the home and the workaday (Graburn 1983a:11–12)

Graburn (1983a:12) further identifies the temporal structure of holiday-ing with the structure of ritual behavior. Both of them have their temporal boundaries through which the sacred is separated from the profane. Holiday-making thus offers people entry into another kind of "moral state in which mental, expressive, and cultural needs come to the fore", and in which "normal 'instrumental' life and the business of making a living" are left behind (1983a:11). Such a moral state has its own temporal boundaries.

Holiday-making relates to time in another three ways. First, as a regular occurrence on an annual or semi-annual basis, holiday-making is an institution. It is an institution by definition because of its regular occurrence. Second, it is temporary and of limited duration—most people cannot be on holiday all the time. Third, it follows a different tempo from the rhythm of home. It is, in a sense, "time off".

More appropriately, holiday-making is an institution of *escape*. Holidaying can be seen as "a mass retreat of square pegs from the round holes of their uncongenial occupations to the square holes in which the best that is in them comes uppermost" (Michael 1950b:14). As a way of escape, holiday-making can be further defined as both "escape from" and "escape to" (Brown 1950). Holiday-makers escape from X to Y (1950:276). For Brown, X mainly consists of the daily environment, daily duties and obligations, daily routines, and daily social relations, while Y consists of those that promise opposite, or different, features in relation to X. "Escape from" and "escape to" can be translated into "freedom from" and "freedom to" respectively. As regards the former, holidaying is a socially and culturally approved way of spending a certain period of time exempt from daily duties and their associated tempos. As Neumann writes:

> People take to the world of travel and nature to flee the routines of work, home, and family and seek out ways of living that involve them in situations where they find they can be closer to some primary and basic mode of life. The river, the trail, and the road are places where the alienating rhythms, routines, and boredom of modern life seem less imposing. They are places where people may find individuality, excitement, flexibility, and freedom (1992:186).

Thus, in more general terms, holiday-making is freedom from the modernized mode of existence that is associated with rigid schedules, deadening routines, and stressful deadlines.

As regards "freedom to", people on holiday have entry into an alternative track of tempos and rhythms. The routines and constraining schedules relating to home are placed in abeyance. Holiday-making thus becomes freedom to change. It follows a metaphorically temporal "rhetoric" that relates to the experiences of change. Holiday-making is a *reorganization of experiences*. It is a cultural construction of alternative temporality.

Besides alternative tempos and rhythms, holidaying involves the freedom to pursue alternative lifestyles, namely holiday lifestyles which are different from the daily forms of life. "Lifestyle" (or "life-style", or "style of life") is a widely used but ambiguous, concept. For Weber (1978) it is used in relation to "status honour". For Simmel (1990) it is akin to "the form of life" and refers to the modern objective form of life, such as lack of character, the calculating feature of modern life, and so on, shaped by the money economy. For Bourdieu (1984) it is linked to the "habitus" and "tastes" of various social classes, which run parallel to various conditions of class existence and economic capital. For Giddens it is related to choice and self-identity. Lifestyle gives "material form to a particular narrative of self-identity" (1991:81). Based on a literature review, Veal defines lifestyle as "the distinctive pattern of personal and social behaviour characteristic of an individual or a group" (1993:247).

According to Veal (1993), there are several points regarding lifestyle which must be taken into account. First, it involves activities, but does not include paid work or occupations. Second, it comprises values and attitudes. However, these are not necessarily part of lifestyle itself; rather they are *influences* on it. Third, whereas lifestyle is often a group phenomenon, it is also an individual matter. A person has a unique lifestyle. Fourth, individuals share a common lifestyle. They do not necessarily have any social contact, although most probably they will. Group inter-action is not therefore a necessary feature of lifestyle, for example a person can share a common lifestyle with remote urban dwellers. Fifth, coherence is likely to be a key variable in analyzing lifestyle. However, it is not a necessary component of the definition of lifestyle, for some life-styles may lack coherence. Sixth, people's perceptions of others' lifestyles are often partial, superficial, and inaccurate. Thus, while recognizability may be a feature of some lifestyles, it is not a necessary part of its defini-tion. Seventh, lifestyle involves choice, and the degree of freedom of choice varies from individual to individual, from group to group, and from time to time. Consequently, the concept of "lifestyle" is different from that of "way of life". The latter is associated with a low degree of choice and the characteristic of "being imposed" (e.g., the situation of poverty or powerlessness), whereas the former is linked to high degrees of free choice (as in the situation of affluence or powerfulness).

Undoubtedly the whole range of life can be stylized, and holiday-mak-ing is one characteristic of a larger lifestyle. Therefore, lifestyles in an industrialized or post-industrialized society include holiday lifestyle as an integral element. Holiday-making itself can also be regarded as a specific lifestyle separate from the routines of daily lifestyle. The regular and circular inversion of holiday lifestyles and daily lifestyles are characteristic of the forms of life under the conditions of modernity and postmoder-nity.

A Phenomenology of Holiday Time

Ryan states that: "holidays do not make sense without reference to non-holiday time" (1997b:200). Indeed, in order to understand the phenom-enology of holiday time, one needs to understand the experience of working time and daily routine. As Saram points out, people have "the need for experiential contrast, to states of boredom. Contrast is achieved primarily in two ways, as deviance, and secondly, what is termed 'time-out' and 'institutionalized evasion'" (1983:92). Greenblat and Gagnon (1983:95) also argue that, with many of the competence-enhancing aspects of life formally associated with work having subsided into routine, people tend to search for such experiences in leisure activities and holi-day experiences. Holiday time is thus a kind of experience of time con-trary to the experience of working time.

In contrast with working time and daily routines, holiday time is an *unusual* time. Voase claims that a holiday is "abnormal" in the sense that "it is infrequent, a break from routine, and arouses expectations of plea-sure and interest which the daily round does not offer" (1995:32). The experience of holiday time is exemplified by Gottlieb's description of American tourists' experience of holiday rhythm in a resort:

> Many Americans like to envision vacations as "time off"—in a sense, the denial of time as one conceives it. One should make love all day, or stay up all night and sleep in the day, denying the "normal" rituals of the temporal sequence. Some vacationers even leave their watches home as a clear symbol of adapting to a non-normal "schedule", eating when they are hungry instead of when the clock tells them to, and so on (1982:170).

Similarly, Michael suggests that, for an ideal holiday, there should be no "arbitrary time-limit" and timetables. People "must have plenty of time, time to stop, to look and to investigate, time to turn aside and follow his nose or his intuition, and above all, the time to linger where delight is" (1950b:5). Although Michael is referring to sightseeing, his comments can also be applied to other kinds of holiday. Of course, the duration of a holiday may be constrained by daily responsibilities and time budgeting. In Ryan's words, "while the holiday-maker may seek to free him or herself from the normal contraints of time, it is done so within a constrained period" (1997b:202). In this sense, holiday time is planned and manipulated time. However, within the temporal boundary of holiday, to achieve an alternative experience of time, namely free time, holidaymakers tend to forget time, deroutinize it, and take a differential lifestyle free from the constraints of regulatory time. Ryan (1997b:201) has identified the experiential time characterizing a holiday as follows:

- Possible freedom from the usual regulatory constraints of time.
- The sense of time is shaped by the vividness of the experience of events.
- Time is a social construct, and thus within holidays is sensed as a consequence of the social interactions that take place.
- Holidays provide the potential to experience time in a way more akin to that associated with indigenous people—that is, time as a natural phenomenon associated with the rhythms of the season, or at least rhythms not imposed by daily work patterns.
- Holidays represent the opportunity to, within a short period of calender time, experience time as other than fixed units of measurement—that is, time seems to speed up or slow down.
- Holidays provide the opportunity to create memorable time.
- Thus holiday time can be used to create positive memories which are assets for the future.
- That time which is a social construct contributes to concepts of self, thereby reinforcing the importance of holidays as periods of potential self-awareness.

Ryan's account of holiday time is phenomenological in nature. This account describes a sense and a kind of experience of time that are both "time off" from daily rhythms and "time for" an alternative experience and existence.

It is suggested here that another feature can be added to holiday time, that is, the consumption of time itself. If in daily life time is something that is to pass, then on holiday it is something for consumption. Holiday-makers are consuming a "good" or "happy" time (this point will be discussed in more detail in Chapter 9). Thus, time on holiday is highly subjective. It is the duration of particular kinds of feeling derived from holiday-making which can not be measured by clock time. The effects of a holiday transcend the moment of holidaying and continue well into the post-holiday period.

Holiday-making is an art-like experience of time and echoes the essence of art. The vitality of art lies in creation, freedom, and change. Once art becomes routine it loses its vitality. Similarly holiday-making is a quest for change and escape from the daily routine. Whereas an artist seeks change and creativity with the aid of imagination, holiday-makers experience change mainly through mobility. Metaphorically speaking, holidaying is an experiential time drama. Holiday-makers are mobile audiences who leave home and routine behind by traveling. "Performed as an art, travel becomes one means of 'worldmaking'" (Adler 1989b:1368). For holiday-makers the world becomes a gradually unfolding picture with the passage of time. The synchrony of the world is translated into the diachrony of itineraries. It is thus through the consumption of holiday time that space, as an unfolding landscape, it also visually consumed. The consumption of holiday time thus implies the consumption of space as well. As a result an individual's relationship with the world changes. The world is no longer a totality beyond the reach of the individual, but rather becomes negotiable, flexible, and accessible when the holiday-maker sets out on a holiday journey. The consumption of holiday time, as a kind of leisure movement, indicates that the world can be experienced as a series of spectacles. Holiday-makers, by consuming time, rebuild a relationship with the world. The world becomes an open, accessible book for them.

In summary, holiday-making is a particular culture of time. It is a social and cultural construction of free time, which is typically understood to be exempt from dominant modern tempos, rhythms, and temporal constraints, on the one hand, and duties, obligations, and the work ethic on the other. It also provides access to alternative experiences of time and lifestyle.

The sociogenesis of modern holiday-making has to do with the emergence of the modern individual's legitimate right to leisure, travel, and holidays with pay. Such rights are gained not merely because of advances of technology or improvements in the standard of living, but are also due to an increased functional interdependency between classes and groups. Holidays with pay are one of the consequences of modern industrialization and industrial relations.

Holiday-making is in a sense a culturally constructed "need" set against the background of the negative side of modern tempos (such as time constraints, pressures of schedules and deadlines, and dulling routines). It is, however, only based on increasing productivity related to the modern temporal order that the negative side of the same temporal order is collectively and culturally defined as "intolerable". Hence, there has been a collective legitimization and naturalization of the "need" for escape and holiday-making as an annual or semi-annual event.

Holiday-making entails a newly emerged social ethic of free time. It is a socially and culturally approved way of spending a certain period of time that is legitimately exempt from the work ethic, obligations, and working tempos. It is both an institutionalized "freedom from" work and everyday rhythms, and a "freedom to" pursue a specific lifestyle contrary to the daily ways of life. A holiday lifestyle involves the consumption of qualitative and differential time. It is an alternative mode of "being-in-the-world", an alternative belonging to the world and others, and an alternative experience of self.

Chapter 6

Modernity and International Tourism

Earlier, reference was made to Cohen's classification of sightseers and vacationers. For Cohen (1974:544–545), whereas vacationers seek change, sightseers seek novelty. Just why and how holidaymakers seek change was discussed in the preceding chapter in terms of time and tempo in modernity. In this chapter, the focus is on why and how sightseers, particularly international sightseers, seek novelty, and this will be examined in relation to the socio-spatial condition of modernity.

Novelty is a visual concept. As Cohen (1974:544) observes, novelty is new only once; once it is seen, it is no longer novel. Novelty-seeking is associated with "curiosity", and curiosity is one of the strongest motives characterizing tourism, particularly sightseeing (Rojek 1993:199). At a certain level, tourist motivation can be explained in terms of this biological or psychological "curiosity" or "novelty-seeking". However, curiosity can be further considered in relation to a wider context in which the drive of curiosity or novelty-seeking is socially and culturally sanctioned and exploited. This wider context is, it is suggested, the socio-spatial condition of modernity. Tourism, as a curious quest for something that is novel and different, involves spaces and people's social relations to spaces, simply because novelty and difference exist not at home but elsewhere.

For tourism to occur, two conditions, among others, must be satisfied. First, since tourists travel from X to Y there must be something in Y that is missing in, or different from, X. To put it another way, for tourists Y is not substitutable by X, as otherwise they would have no reason to travel to Y: they could just remain in X. Second, there should be no spatial or social obstacle on the way to Y. In other words, Y must be accessible, as otherwise, tourists may choose to remain at home or travel to destinations that are accessible. For example, if there are territorial border closures, xenophobia, or wars in destination Y, then, Y is neither spatially

117

nor socially accessible. Both of these two conditions involve a connection between the social factor and the spatial factor. On the one hand, for tourists in X the travel to Y involves a *spatial* factor. First, Y, as a space of culture, society, or physical environment, must have something in qualitative contrast to X. This is what Saram calls the "qualitative acceptability" of tourist regions (1983:92). Second, they must be able to eliminate the space or distance between X and Y, and have access to Y. On the other hand, travel to Y also involves a *social* factor. First, Y must be culturally, socially, and even politically acceptable to tourists in X. Second, Y must not intend to exclude tourists from its territory, and must be willing to receive and host incoming visitors.

If "novelty" is a psychological and visual term, then, the sociological equivalent is "difference", which refers both to spatial differences, such as natural features of the physical environment of other places, and to socio-cultural differences, such as exotic cultures. Thus, "difference" highlights a particular social–spatial nexus. Within the context of modernity, tourism, particularly international tourism, can be regarded as a culturally and socially acceptable and encouraged "celebration of difference" (MacCannell 1976; Van den Abbeele 1980). For Northern and Western tourists, spatial differences are exemplified by the pleasant weather in the Mediterranean coastal resorts, and socio-cultural differences are exhibited in oriental or primitive cultures. "Difference" is a sociological term because the spatial and socio-cultural differences to which "difference" refers can be understood sociologically and treated as the result of social construction. "Difference" is a sociological demarcation between the familiar and the unfamiliar, the usual and the unusual, the ordinary and the extraordinary, routine and novelty, "us" and "them", through which social reality and social identities are constructed. Thus, as well as treating tourism as novelty-seeking in a psychological sense, tourism can be regarded as difference-seeking in a sociological sense. Similarly, while novelty-seeking is seen as motivated by innate curiosity from a psychological standpoint, from a sociological perspective difference-seeking is regarded as positively sanctioned by certain values, attitudes, and images held in society about the spatial and socio-cultural differences that exist elsewhere. Under the condition of modernity, people have developed a broader cultural interest in these differences, a process closely related to the formation of national and social identity. This cultural interest in differences will be treated here as a *cultural curiosity* (as distinguished from psychological curiosity) for differences. Hence, this chapter will treat international tourism as difference-seeking and discuss international tourism in relation to the social–spatial nexus of modernity.

This chapter consists of four sections. The first analyzes the spatial–social nexus relating to nationalization and globalization under modernity. The second examines the relationship between cultural curiosity and nationalization, and hence the relationship between nationalization and international tourism. The third explores the relationship between globalization and the exoticism which characterizes international tourism. In the final section, international tourism will be discussed in relation to the social construction of the space of differences, in which sightseeing takes place. It also involves the problem of the management of novelty and unfamiliarity since tourists' taken-for-granted skills of self-management are challenged in the space of strangers.

Modernity and the Social-Spatial Action

Modernity as a social formation cannot be severed from its relationships to spaces or spatial characteristics (Featherstone, Lash and Robertson 1995; Friedland and Boden 1994; Giddens 1990; Harvey 1990; Shields 1991; Soja 1989; Urry 1995), just as it cannot be detached from its relationship to time and temporal order, as discussed in Chapter 5. What are the relationships between modernity and spaces? Before this question is answered, the meaning of space and spatial action needs to be clarified.

Spaces and Social–Spatial Actions

"Space" is an age-old concept in philosophy. Roughly speaking there are three different philosophical conceptions of space. First, the absolutist conception of space is that space is the extension of corporeal substance (Descartes) and remains similar and immovable (Newton). Second, the relationist conception of space is the idea that space is nothing but the spatial relations of substances in an order of coexistence (Leibniz). Third, there is the *a priorist* conception of space, which, in a sense, is also an absolutist conception. Here it is proposed that space is an *a priori* category whereby human beings organize and structure their sensual experiences (Kant). Philosophical concepts of space are on the most abstract level. They are about *space as such* rather than about concrete spaces in commonsense. They abstract out various concrete social contents such as asymmetrical powers and so on from the concept of space.

When applied to the social sciences, the philosophical concept of space "as such" becomes less relevant, for spaces are not something given, but rather the result of social construction and contestation. From a phenomenological perspective, space is always defined in relation to human beings' experiences. Persons treat themselves as the centres from which distance and proximity, vastness and smallness, right and left, front and back, and so on are structured (Tuan 1977). From a sociological perspective, spaces always presuppose the question "whose spaces are they?" In this sense, spaces are the outcome of certain social arrangements, rather than vice versa. Usually "space" is employed in the social sciences in the following senses. First, it is is used as "a frame of reference for actions" (Werlen 1993:3). This involves the questions "where are we?" and "how big a space do we have?" Space thus acts as an empirical spatial framework through which the spatial locations of human beings and areas for social activities are defined. Second, space is used in relation to time and regarded as spatial extension. It is thus treated as an integral element of social structuration (Giddens 1979, 1981, 1984). It is also seen as a changeable dimension in the time–space nexus (Friedland and Boden 1994) or in "time–space compression" (Harvey 1990). Third, space is treated as a metaphor for opportunities (e.g., space of profit, space of employment), significance (e.g., space of welfare, space of justice), social approval (e.g., there is no space for drug-taking, space of privacy), visual representation of the invisible "black box" phenomenon (e.g., mental space, cognitive space, logic space) or life chance (space of minorities, space of underclass). Metaphorically, space is far from being physical and three-dimensional. Nor is it a space "as such" in a philosophical sense. Finally, space can be regarded as a geographical spatial area, spatial distance, and territories in which spatial relations and arrangements of society constitute the social–spatial nexus (in certain circumstances space is also considered as akin to place). In this sense, space is seen as containing social contents and involves social relationships. Most social geographers and sociologists seem to use space in this sense (Castells 1977; Entrikin 1991; Foucault 1977; Gregory and Urry 1985; Harvey 1985; Lefebvre 1991; Poulantzas 1978; Sack 1992; Saunders 1985, 1989; Shields 1991, 1992; Soja 1989; Urry 1985). Thus, Urry claims that, "there is really no 'space' as such—only different spaces" (1991b:174) while Sayer argues that, "space makes a difference, but only in terms of the particular causal powers and liabilities constituting it" (1985:52). Space in this sense involves spatial relations which are closely related to social relations in society. Generally speaking, spaces are the effects of social actions, social powers, or social processes, rather than their causes (Werlen 1993), though they have relative autonomy and constitute conditions for further social actions. In addition, there has recently emerged a new concept: "hyperspace", "cyberspace", and "virtual space" refer to electronic space such as the Internet and CD-ROM.

In association with social action, there is also *spatial* action, or *space-oriented* action. For Weber, "Action is 'social' insofar as its subjective meaning takes account of the behavior of others and is thereby oriented in its course" (1978:4). Similarly, action is spatial insofar as its subjective meaning takes account of spaces or spatial conditions, and is thereby also oriented in its course. Spatial action and social action are only distinguishable conceptually; in reality they coincide with or overlap one another. In this sense, just as one can speak of "social action", one may also refer to "spatial action" or "space-oriented action". As spatial action necessarily involves taking account of others and is thereby oriented in its course, spatial action is also social, that is, "social–spatial action". To paraphrase Weber, action is social–spatial insofar as its subjective meaning takes account of both the behavior of others and the spatial conditions, and is thereby oriented in its course. Indeed social interactions and relations may be space-oriented and space-related. Social interactions may involve spatial strategies, such as approaching and distancing, inclusion and exclusion, openness and insulation, aggressiveness and avoidance, occupying and withdrawing, separation and uniting, centralization and marginalization, presence and absence, making-room-for and making-no-room-for, disclosure and enclosure. Heidegger has expressed the idea of space-related strategy by talking about "de-severance", as "making the farness vanish—that is, making the remoteness of something disappear, bringing it close"; and "directionality" as "a direction towards a region out of which what is de-severed brings itself close" (1962:139,143). Schutz and Luckmann have also revealed the spatial aspect of social activity by making a distinction between "the primary zone of operation" and "the secondary zone of operation" (1974:44). Sack has succinctly summarized the importance of space to action: "Space is ontologically fundamental to agency, in that it is part of action, and constitutes the experience of being in the world" (1992:19).

Social–spatial action will bring about intended or unintended spatial consequences—either the reproduction of a given social–spatial nexus or the transformation of their nexus. This raises the question of whether there is a certain social–spatial nexus distinctive to modernity, or, to put it another way, what is the ideal-typical social–spatial action characteristic of modernity? According to Weber (1978:24–5) there are four ideal types of action: purposive-rational action, value-rational action, affective (emotional) action, and traditional action. Correspondingly there are also four ideal types of social–spatial action: purposive-rational (such as urban and town planning), value-rational (such as pilgrimage and sacralization of the national territory), affective (such as home or place attachment), and traditional (such as choosing the location of a house or tomb in the light of the geomancer's judgements in traditional China). It is hypothesized here that modernity is characterized by both purposive-rational and value-rational social–spatial action (or social–spatial nexus).

It is quite evident that the space that each nation-state society occupies is limited. Each nation-state is located in a specific spatial position and within a restricted area. Clearly "where we are" is not an essential feature that distinguishes modernity from premodernity. Rather, ideal-typically speaking, what makes a modern society different from traditional society in terms of the social–spatial nexus, is that traditional society is highly constrained by the limits of the physical space in which it is located. By contrast, modern society is able to transcend the boundaries of physical space or spatial conditions in terms of rational spatial strategies.

Under modernity this transcendence of the limits of physical space consists of two interrelated processes: the symbolic and cultural transformation of space or territory, and the economic rationalization of space. The latter involves purposive-rational social–spatial action, and the former involves value-rational social–spatial action.

The purposive rationality of social–spatial action is exhibited in the efficiency of achieving a given goal with the least spatial effort. This is exemplified in two different dimensions: the spatial efficiency of social transaction, and the spatial efficiency of social production. Social *transaction* involves social–spatial interaction and spatial arrangements between people, whereas social *production* involves spatial arrangements between people, and between people and physical resources. Social production and social transaction are dialectically intertwined with and penetrated by each other.

The rationality of social–spatial action relating to social *production* in modernity is embodied in the most efficient distribution of, access to, and use of resources. The limits of physical space are transcended by the modern technologies of transportation and communication, and by the modern market economy and commodity production systems. The modern market economy entails a *spatial* as well as a *social* division of labor on which the social–spatial integration of social production depends. One of the implications of social–spatial integration in social production, based on the social–spatial division of labor, is that modernity witnesses an increasing interpenetration of presence and absence (system integration across space and time, separated from face-to-face social integration) (Friedland and Boden 1994; Giddens 1981; Shields 1992). Here all social–spatial barriers, such as distance and border control, are in principle removable. Under modernity, capital entails the ability to remove any social–spatial barrier (Harvey 1990). As a result, regionalization and globalization arise as social–spatial processes which aim to transcend the limit of resources in the space of a local or national economy. However, as history shows, regionalization and globalization are not without tensions and conflicts.

The rationality of social–spatial action relating to social *transaction* in modernity is typically shown in the urban form of social–spatial strategies employed by people and social–spatial relations between people. The urbanization of populations was to a large extent linked to industrialization in the process of modernization. However, the urbanization of the population itself raises the issue of how to conduct social transactions in an efficient way, since the concentration of populations (urbanization) bankrupts the traditional form of "dense sociality" or social transaction in communal ways (Bauman 1987). Various rational social–spatial strategies thus occur in urban contexts of physical proximity and crowd: withdrawal from overstimulation (Simmel 1990); distancing relationships of neighbourhood, civil inattention (Giddens 1989); footloose labor forces and the associated low attachment to place (Seamon 1979); low commitment to anonymous people, spatial segregation and segmentation (Sack 1986); territorial inaccessibility to personal and private space (back region) (Goffman 1959); and so on. Among other things, it is the separation of *public* and *private* social space that is most significant. People tend to limit their "dense transaction" to private space (intimacy), while in public space a strategy of standardized, superficial, or "heartless" social transaction is adopted.

However, the purposive-rational social–spatial action that is characteristic of modernity is often contradictory. To maximize national interests, various national actors who employ a purposive-rational social–spatial strategy tend to expand their spaces, and such expansion sometimes leads to imperialist conflicts and wars between Western nations in captured colonies. Thus, rationality ends up as irrationality. The way in which purposive–rational social–spatial action leads to irrationality is a complicated issue, but clearly it has to do with the sentiment of nationalism, a feature characterizing value-rational social–spatial action.

A distinctive social–spatial feature of modernity is the formation of the nation-state with its own national territory. This involves the value-rational transformation of space into the territory of the nation-state that is called by Anderson (1983) the "imagined community". Consequently, a myth of the nation-state is formed and its space is sacralized. In association with this a national identity is formed and the sentiment of nationalism is developed. This helps to bind together the members of the nation.

With the advent of industrial capitalism, the traditional attachment to place (usually birthplace) is weakened. People become footloose, mobile, and "opportunity-oriented" rather than "place-oriented". The first wave of mass emigration of the population from rural areas to the industrial cities in the eighteenth- and nighteenth-century Britain is one such example. The space of industrial capitalism is related to the so-called "abstractization" of space in Lefebvre's (1991) sense. However, this "abstractization" of space does not prevent people from having affective attachment to the symbolic space of the nation-state as a whole. Rather, the formation of sentiments of nationalism and patriotism, in the process of sacralization of the symbolic space of the nation, is a crucial element in the formation of national identity under modernity. Therefore, geographical spaces are not only purposive-rationally transformed in the economic dimension (economic exchange between places or countries and economic globalization) and in the political dimension (sovereignty and territory), but also value-rationally transformed in the symbolic and cultural dimension (nationalism and national identity). If the economic rationalization of space helps to overcome the physical limits of a nation-state's physical spaces or resources, then the symbolic and cultural transformation of space helps the people of the nation to hold to this space in their "hearts"; thus, no matter where they go they have affective attachment to their "motherland".

Nationalization and Globalization as Two Social–Spatial Strategies

Geographers speak of "territorial action" and use "territoriality" to refer to the locus in which social relations come to join spatial relations (Sack 1986, 1992:19). Territorial action can be regarded as social–spatial action and territoriality as the spatial result of social relationships, namely a social space or a social–spatial nexus. Society always involves social–spatial actions regarding spaces or territoriality. A certain social–spatial action leads to a corresponding social–spatial nexus, namely a spatial result (spatial relation and arrangement) of powers and social relations. As a kind of social–spatial action or strategy, the agents of power always tend to include something and exclude others, centralize something and marginalize others, approach something and distance others, make something present and make others absent, within a certain social space. Thus, Shields suggests that three spatial forms – inclusion and exclusion, differentiation, and presence and absence - are particularly significant: they "operate together to underwrite the modernist 'social spatialization' of Western cultures" (1992:184).

In relation to modernity, two interrelated social–spatial strategies can be identified. First, *nationalization* can be regarded as a kind of social–spatial strategy that leads to a cultural and symbolic transformation of national space or territoriality. Nationalization involves not only the formation of the sentiment of nationalism, but also the creation of special sacred places or territories in the myths and traditions of a nation. These may result in certain national symbols, such as monuments, through which national identity is communicated. These symbolic spaces become tourist sites for domestic or foreign tourists in late modernity. However, to the extent that nationalization is both a defensive and a territorial planning strategy, it also includes purposive-rational social–spatial actions. It involves a process of rationalization of national space in which dangerous otherness, enemies, strangerhood, and difference are socio-spatially excluded, blocked or even conquered. In this way it tries to create a safe and secure space for the majority of domestic citizens. Second, there is globalization or denationalization (Featherstone 1990;

Giddens 1990; King 1991; Robertson 1992; Waters 1995). Globalization is a process of transcending national territorial boundaries and of approaching the world as a whole. Although globalization may precede modernization (Robertson 1992), it is in modernity, particularly late modernity, that globalization is accelerated. Modernity entails globalization. In its early stage it appears as imperialism and colonialism. In its late stage it is demonstrated in the increasing "villagization" of the Earth. As Wallerstein explains, "Capitalism was from the beginning an affair of the world economy and not of nation-states. . ... Capital has never allowed its aspirations to be determined by national boundaries" (1979:19).

Nationalization, Differences, and Cultural Curiosity

Nationalization and globalization are interconnected social–spatial processes in modernity (Arnason 1990:224–5; Giddens 1990) and both are related to international tourism. The relationship between nationalization and international tourism is discussed here, and that between globalization and international tourism will be touched on in next section.

Nationalization and the Spatial Politics of Differences

Modernity, it is argued, involves the "nationalization of social practices" (Wagner 1994:86) within the territory of the nation-state. The social process of nationalization and the spatial process of territorization are two sides of the same coin. According to Giddens (1984:164), modern society is in fact the nation-state. In other words, in the context of modernity there is no "society" as such, but only societies of the nation-state. The "emergence of the nation-state was integrally bound up with the expansion of capitalism" (Giddens 1981:12). With the advent of capitalism the territorially bounded nation-state replaced the city as a "power-container" (1981:12, 147).

In the West, the nation-state emerged as a more powerful agent of order and control when the traditional communal forms—via the form of "mutual gaze" and "dense sociality"—was bankrupted (Bauman 1987). In addition to economic take-off based on the techniques of capital accumulation, political take-off was ensured by the nation-state, by inventing the social–spatial techniques of administering the accumulation of population (Foucault 1977:220–221). The nation-state was one of the most powerful agents for territorialized nationalization and universalization. Under modernity, particularly early modernity, the nation-state was characterized by a modern spatial politics of difference. In Bauman's (1987, 1990a) terms, it was characterized by its cultural intolerance of, or impatience with, differences and non-conformity, such as ethnic, linguistic, religious, legal, political, or cultural "difference". Wars and crusades were frequently waged by the nation-state in order to conquer differences and non-conformity within a territory.

Under early modernity, closely united with the power of the nation-state, intellectuals regarded themselves as "legislators", i.e., the most authoritative arbitrators of opinions due to their holding the "universal" and "absolute" criteria of truth, goodness, and beauty (Bauman 1987). Thus, the power-knowledge nexus tended to endorse a spatial politics of difference. In the minds of these "legislators", differences within that were at odds with the rational order, such as irrationality, the barbarous, the uncivilized, and so on, needed to be wiped out, or at least marginalized (Bauman 1987); and differences from without, if dangerous and in the way of the progress of Western rationality, such as the barbarous Others, had to be conquered or excluded. Therefore, the formation of the nation-state tended to lead to the nationalization of culture based on a "core" ethnicity, and hence produced a cultural space of national "family affinity" (in a Wittgensteinian sense).

However, such attempts at homogenization and universalization of the nation-state by "legislators" (Bauman 1987), and liberals or rationalists in the West, have never been completely successful (Bauman 1990a). For example, "ethnic revival" or "human retribalization" has for a long time acted as a powerful counterforce (Smith 1981). Nevertheless, the tendency of nationalization is to incorporate ethnic or cultural diversity and difference into a hegemonic national character.

This does not imply that there is no "variety" in national culture or national character. Rather, "variety" itself constitutes the spectrum of that national identity or character. However, once "variety" is incorporated into a national culture it cannot be treated as difference or otherness. "Difference", on the other hand, refers to something that cannot be assimilated and incorporated into the spectrum of national identity in the West. From the standpoint of a Western nation, the category "difference" refers to those characteristics that are exhibited in *other nations* or *other cultures*, those that cannot be integrated into its own framework of national identity. "Difference" can, however, be further sub-divided into two types. The first refers to strangeness, otherness, or alienness that is at the odds with, or in the way of, Western rational order, and is hence excluded, banned or separated. In this sense, "difference" presupposes the social–spatial action of exclusion, and implicitly "justifies" imperialists' conquest of others. Thus, "difference" has a particular referent. In Western eyes, phenomena that cannot be in conformity with Western rational order, or those cultures that are regarded as alien to Western culture, for example Oriental or primitive cultures, fall into this category of "difference". The second kind of "difference" refers to neutral pluralism and diversity. In the eyes of a Western nation, for instance, other European or Western national cultures which share some cultural affinity or similarity with it, fall into this category, as do different natural phenomena because they are "neutral" in comparison with "social and cultural differences". In the following discussion "difference" will be used in both the first and the second sense. However, it is the first sense to which particular reference will be made.

Nationalization or the formation of a nation-state therefore involves not only a "place myth" (the cultural and symbolic transformation of space), but also a process of collective definition and representation of differences beyond the nation-state boundaries. A nation or a national identity is formed not by its similarity to other nations, but by its difference from other nations. In this sense difference is a necessary boundary to distinguish one nation from others, and hence helps define the universe of a nation or national identity. Thus the formation of a national identity and the definition (or representation) of difference constitute a dialectical process. A nation's attitude towards difference can be ambivalent. Differences might be seen as a threat or a potential danger, but they might also be regarded as objects of cultural curiosity. How they become the latter involves a complex socio-cultural process.

Cultural Curiosity in International Tourism

Tourism is, in a sense, a satisfaction of cultural curiosity. As Rojek puts it, "Curiosity was one of the strongest motives behind travel for pleasure" (1993:199). Curiosity is a biological and psychological phenomenon which has been well documented by psychologists (Berlyne 1966a, 1966b; James 1890; Spielberger and Starr 1994; Zuckerman 1979). From their perspective, curiosity can be simply defined as a biological or psychological drive to explore novelty, strangeness, or diversity (Spielberger and Starr 1994). Stagl succinctly summarizes five characteristics of psychological curiosity as follows: it is "a directed activity involving locomotion and the senses"; it "has something to do with new or unknown situations"; it is a "'superfluous' activity having no immediate utilitarian goal"; it is "closely connected with play"; and it "leads to indirect, long-range advantages in the form of learning" (1995:2). Curiosity is functional to human beings' survival because it urges them to explore unknown situations. To survive the world, humans must be "world-open", and curiosity is a feature of this world-openness. Curiosity and exploration are so closely linked that they are considered almost identical. Although both characteristics appear in higher animals, such as other mammals or birds, it is in humans that they are most outstandingly exhibited (1995:2). Humans differ from other "curiosity animals" through their use of language, which allows them to enlarge their field of explortion (1995:3). Berlyne (1960) distinguishes between "perceptive curiosity", the direct sensory exploration of unknown objects and situations, and "epistemological curiosity", indirect exploration via the asking of questions and thinking. Humans can transcend curiosity animals by their epistemological curiosity.

Curiosity is also referred to by philosophers. For example Aristotle regards it as the "desire to see" which is essential to cognition: "cognition was conceived in terms of 'desire to see'" (Heidegger 1962:215). For Heidegger, "Primordial and genuine truth lies in pure beholding. This thesis has remained the foundation of Western philosophy ever since. The Hegelian dialectic found in it its motivating conception, and is possible only on the basis of it" (1962:215). Heidegger also treats curiosity as the tendency towards "seeing", as "the tendency towards a peculiar way of letting the world be encountered by us in perception" (1962:214). Furthermore, he elevates curiosity to "existential-ontological" status (1962:214). For pre-Christian philosophers, curiosity is classified into two types: *bona curiositas* and *mala curiositas*. The latter involves the fact that humans "inquire into a neighbor's business, to be hungry for hearing and repeating news, tales, gossip, and novelties, to be ever searching out the ugly and abnormal" (Zacher 1976:19). In contrast with *mala curiositas*, which leads humans to search for a sensitive knowledge, *bona curiositas* refers to the desire to seek intellectual knowledge about new or unknown situations. *Mala curiositas* is always a derogatory term, while *bona curiositas* may be a commendatory term for some but not always for others as different cultures may have different values and attitudes towards it (Zacher 1976).

Curiosity can also be looked at from a sociological perspective. True, it is a biological or psychological phenomenon. However, people's attitudes towards it and the way it is treated are sociological. A sociological perspective on curiosity is different from a psychological perspective in that the former treats it in the following three ways. First, a sociological treatment of curiosity is linked to the issue of how it is sanctioned within a given society or culture. Different cultures have different attitudes, values and norms in respect of curiosity. As a result it may be either negatively or positively sanctioned, constrained or encouraged. Second, with the support of the growth of powers, such as political, economic, military, or intellectual powers, curiosity about a physical or a human object, may be encouraged and cultivated in a society by the simultaneous removal of *fear* of that object. Third, curiosity may be socially and culturally exploited by a society or a group of people to the advantage of certain utilitarian goals, such as scientific discovery or the development of a tourism economy.

In different societies and eras there have been varying attitudes towards curiosity. Throughout the Middle Ages it was regarded as temptation, a vice, a sin, "characterized by a fastidious, excessive, morally diverting interest in things and people" (Zacher 1976:20). Such a view was clearly demonstrated in Aquinas's Work. Curiosity was thus seen by theologians as the antithesis of pilgrimage and as a vice leading the individual away from God. It was also regarded as "a phase of original sin, which made all men wanderers in the fallen world" (1976:4). By contrast, pilgrimage "allowed men to journey through this present world visiting sacred landscapes as long as they keep their gaze permanently fixed on the invisible world beyond" (1976:4). Yet from the fourteenth century onwards, curiosity slowly gained acceptance. Pilgrimage (the devotional practice) and curiosity (the vice) "had become closely identified with one another" (1976:4). Changing attitudes towards travel and curiosity, as exhibited in the work of de Bury (*Philobiblon*), Mandeville (*Travels*) and Chaucer (*Canterbury Tales*), had to do with the origin of modern humanism in the United Kingdom (1976:12). "The interest of the humanists in the external world as well as their enthusiasm for travel were epiphenomena of this change of mentality which prepared for the 'age of discoveries'" (Stagl 1995:49). With the advent of modernity the range of objects about which people were curious was greatly enlarged, and many objects which were previously feared, became objects of interest. This had much to do with the growth of the economic, intellectual and military power of industrial society, which helped remove the previous fear of things and peoples. Thus, for instance, while thunder once caused fear in the hearts of men and women, it may trigger off curiosity in modern people. Furthermore, curiosity can be commercially exploited, for example in marketing travel and tourism under modernity since the time of Thomas Cook. Curiosity is also cultivated in education to encourage intellectual and scientific inquiry.

Thus, if psychology can explain tourist motivation in terms of psychological curiosity, then sociology does so in terms of the attitudes and values that people hold about curiosity and the ways in which they deal with it. Instead of speaking of the subject in a psychological sense, sociology can speak of "cultural curiosity", namely curiosity that is acceptable to and positively sanctioned by a culture. Tourism, then, is bound up with cultural curiosity. Tourism, particularly international tourism, involves the social organization of objects and cultures for the satisfaction of cultural curiosity, based on the exclusion of fear and anxiety regarding "Others" and differences.

As previously mentioned, curiosity is a drive towards novelty-seeking (Cohen 1974; Lee and Crompton 1992). It is a banal observation that novelty must be sought beyond the familiar daily environment. This is perhaps one of the simplest reasons for tourism. Novelty may involve distance, for the farther away a thing is, the more novel it will probably be. However, distance is not a sufficient condition. Certain nearby things may be novel because they are closed to the public. For example, Buckingham Palace is near Londoners, but it is novel for most of them because they have never entered it. That is why so many people visited Buckingham Palace during the summer of 1993, when it was opened to the public by Queen Elizabeth in order to raise funds for the repair of Windsor Castle, which had been damaged by fire. But more usually, novelty exists elsewhere, and mostly abroad, since people's everyday activities are usually limited to their daily or domestic environment. Thus. international tourism, particularly in the form of sightseeing, involves curiosity about other cultures and different ways of life beyond one's own horizon. Cultural curiosity embodies "generalised interest" (Cohen 1972:165) or "a widening circle of interest" (Simmel 1990:76) in other cultures, peoples and associated environments. Thus defined, it is hypothesized here that there is a close link between nationalization and the genesis of cultural curiosity in international tourism.

As discussed above, nationalization tends to build a relatively homogeneous national identity in its "core", based on local diversity and variety. In so doing, nationalization has socio-culturally and socio-politicallly constructed differences that exist beyond its territorial boundaries, as the limits by means of which its universe is defined and its distinctiveness is made clear. Difference is the backdrop against which national identity is put into relief. In this way, territorial nationalization brings about a potential space for curiosity about differences existing elsewhere, about the identities of other nations or ethnicity, or about how foreign or alien peoples live their lives. Territorialized nationalization, while successfully incorporating domestic diversity and variety into the spectrum of national identity, tends to provoke curiosity about difference *beyond* the boundaries of the territory. In this sense, territorialized nationalization is an implicit but necessary condition for the rise of international tourism. In more general terms, nationalized modernity in the West produced a potential space for exploration, travel, and tourism because it created a potential space for cultural curiosity about unfamiliar features beyond the borders of the nation-state. However, such a potential space becomes a reality only when other necessary conditions have been satisfied.

Cultural curiosity is an interest in and an impulse to explore the differences exhibited in other cultures. However, cultural differences do not automatically and necessarily become the object of curiosity and interest. In certain circumstances they instead become the object of fear and anxiety. As mentioned above, from an occidental perspective, "difference" specifically refers to non-assimilatable, non-Western, non-Christian, and non-rational "Otherness" and strangeness. Such a difference' may often be the object of fear and anxiety (Morley and Robins

1995). Fear is possibly a shadow of curiosity. Too much fear of a person or object generates a "fight or flight" attitude, and renders the satisfaction of curiosity impossible. Curiosity is generated only when that fear is overcome. In respect of tourism, cultural interest in Others and differences implies that there is no shock, threat, or danger in encounters with Others and differences. As Boniface and Fowler put it, "Difference has a place, but for many people the crucial element in a touristic experience is that it should not threaten, or allow them to feel uncomfortably deprived of the comforts of home" (1993:7). Thus, there is a zero-sum relationship in the tension between curiosity and fear. To make this relationship favourable to curiosity, many conditions need to be satisfied. Above all, a "cultural self-confidence" is needed (Graburn 1983a).

Nationalization seems to play a positive role in the building of national or cultural self-confidence in relation to differences. As a result, nationalization gives rise to a sacralization of an "imagined community" and national identity. National identity is then a source of pride and self-confidence. It is therefore a spiritual support for curiosity about difference, and helps to overcome fear of differences and "Otherness". Without a national identity of which people can feel proud, individuals may either avoid Others due to fear, or simply subject themselves to the ruling of Others (such as Friday subjecting himself to the ruling of Robinson Crusoe). Thus, in early modernity Britain was the first and most important country to send tourists abroad, partly because of its advanced industrial civilization and associated national self-confidence and pride. As a popular guidebook written by Abraham Eldon in 1828 stated:

> Let every man who leaves England convince himself of one thing—that he will see nothing like it until his return. England was, is, and always shall be the envy of surrounding nations (quoted in Swinglehurst 1974:40).

National self-confidence associated with national identity is strong only when it is supported by strong economic, political, and military powers in relation to other nations. In this sense, it is not nationalization alone, but the interweaving of nationalization and modernization that brings about a sense of superiority in national identity, and hence a strong cultural self-confidence. National identity triumphs particularly when a nation can defeat its "enemies" either militarily or—as in the case of Germany and Japan, where this strategy ultimately failed—in economic terms, whenever confrontation takes place between a nation and its "enemies".

Therefore, nationalization, together with modernization, brings about a sacralized national identity, and a strong national or cultural self-confidence which removes the fear of the difference of Others. Thus, far from avoiding Others, Western modernity witnessed increasing encounters with Others and differences beyond the boundaries of national territories. However, in *early* modernity, when empire and imperialism were an important stage in Western national development, the approach to Others and Otherness was to a large extent based on negative images of Otherness as something standing in the way of progress, and should therefore be overcome. The aftermath of this situation in the post-colonial period is a combination of old nostalgia and new curiosity. This post-colonial cultural curiosity is related to a new wave of globalization in the post-war period.

Globalization, Mediatized Images, and Exoticism

Giddens argues that "the advent of modern capitalism is concentrated upon the tension between the internationalizing of capital (and of capitalistic mechanisms as a whole) and the internal consolidation of nation-states" (1984:197–198). Nationalization is intertwined with globalization. If nationalization tends to exclude differences within, then globalization has tended to approach and deal with differences from its inception. In the age of colonialism, globalization stemming from the West tended to wipe out or conquer the differences and Others that existed in the colonial destinations, as shown in the massacre of indigenous American Indians by European whites. In this sense, globalization was an accomplice of ethnocentric nationalization in the eradication of difference. This "crude colonialist globalization" no longer survives and most colonies have assumed independence, though economically weak nations still depend to some extent, upon the economically advanced nations. However, "difference" and Others, in the form of immigrants, penetrated West after World War Two and globalization began to acquire its "neutral" meaning. Globalization refers to "A social process in which the constraints of geography on social and cultural arrangements recede and in which people become increasingly aware that they are receding" (Waters 1995:3, emphasis deleted).

The World of Diversity

Theoretically, globalization is the increasing denationalization of certain phenomena such as capital, finance, technologies, communications, management skills, commodities (e.g., Coca Cola), lifestyles, information, ideas, values, and beliefs. It would be foolhardy to conclude that the nation-state is bankrupt and that the cultural consequences of nationalization (e.g., national identity) are obsolete. In fact "the emergence of the nation-state is itself a product of globalization processes" (Waters 1995:98). Thus, the point at issue is that, in certain aspects, the power of the nation-state over certain affairs within its territory has been weakened or undermined since it is subject to the political power of international organizations (e.g., the UN, the EU) and to the economic power of trans- and multinational corporations. But in other areas the sovereignty of the nation-state remains and the sense of national identity has been reinforced, rather than weakened. Thus, although globalization is to a certain extent an homogenizing process, it is not synonymous with homogenization (Appadurai 1990:307; Arnason 1990:224; Featherstone 1991a:147; Hannerz 1990:237). More correctly, it is a differentiating as well as an homogenizing process (Waters 1995). In regard to cultures, globalization leads to an increase in cultural interaction, cultural contact, and "cultural creolization" other than cultural homogenization (Hannerz 1987, 1990). As Featherstone states:

> the globalization process should be regarded as opening up the sense that now the world is a single place with increased contact becoming unavoidable, we necessarily have greater dialogue between various nation-states, blocs and civilizations: a dialogical space in which we can expect a good deal of disagreement, clashing of perspectives and conflict, not just working together and consensus (1995:102).

Globalization does not eradicate difference, variety, and diversity. Historically, the Enlightenment thinkers insisted that globalization, as the extension of Western order, reason, values, and civilization to the rest of the world, should lead the universalization, homogenization, and Westernization of the world. History has proved that this was not only a justification of colonialism, but also a utopia. In reality, even if globalization has indeed led to what Hannerz (1987) calls "cultural creolization", it has not eradicated difference. Rather, difference is alive and well. The world today is one of difference, diversity, and pluralism rather than sameness.

Not only does globalization fail to remove difference and diversity; it also provides easy access to them because it has transformed the Earth into a "global village", a term coined by Marshall McLuhan (1964) to refer to a changing sense of the time-space of the world, brought about by advances in media technology. Nowadays many more terms confirm the sense of there being a "global village", such as "time-space compression" (Harvey 1990) and "the interpenetration of presence and absence" (Giddens 1990). Waters argues that "McLuhan's global village was perhaps misnamed because a village without circuits of gossip would be strange indeed" (1995:150). Such globalized circuits of gossip are now becoming possible by means of today's telecommunication systems.

Indeed, people's experience of space and time, a sense of the here and now, has been dramatically changed by the effects of contemporary globalization. Friedland and Boden describe this situation well;

Modernity has ... brought enormous and increasing changes in the tensions between the immediacy of here and now, our physical location in space and time, and the sorts of experiences, actions, events, and whole worlds in which we can partake at a distance. Our experience of here and now has increasingly lost its immediate spatiotemporal referents and has become tied to and contingent on actors and actions at a distance. The experiential here and now of modernity is thus in a real sense nowhere yet everywhere (1994:6).

It is quite obvious that such experience of the here and now has increasingly changed people's orientations and perceptions, and their image of the world as a whole. If in early modernity a trip for the sake of curiosity to a remote destination was unlikely because of the high cost of time as well as the high risks, nowadays it is simply seen as normal. The greatly improved communications, transportation, and rapport between most nations, has broadened the "comparative interaction of different forms of life" (Robertson 1992:27). Indeed, the world has become so small that the traditional idea of presence no longer applies. People can be in London today and Tokyo tomorrow, a presence unthinkable before. Foreign countries have come within the reach of "global villagers": people who enjoy tourism and wish to satisfy their cultural curiosity (Boniface and Fowler 1993). Tourist destinations have become an extension of home, and foreign countries have become what Hannerz (restating Paul Theroux's view) calls "home plus": "Spain is home plus sunshine, India is home plus servants, Africa is home plus elephants and lions" (Hannerz 1990:241).

International tourism, which has thrived during the post-war period, is the result of globalization. From the 1970s onwards, airlines, tour companies, hotels, marketing, insurance, and banking, all of which are essential to international tourism, have increasingly become internationalized, transnationalized, or multinationalized (Lanfant 1980; Shaw and Williams 1994). They are, to a large extent, globalized.

Therefore, although globalization is in tension with nationalization, it does not wipe out difference and diversity in the world. Globalization does not lead to the elimination of international objects of curiosity (differences). Conversely, globalization does lead to the "villagization" of the world, and thus changes people's sense of time-space and provides them with easier access to difference and diversity. In this sense, globalization, particularly in the post-war period, is helpful to the development of international tourism.

Globalization and the Mediatization of Images of Difference

The ways in which globalization influences international tourism are numerous. For example, it helps to overcome fears about the Other, to increase knowledge about other nations based on increasing interaction, to enhance international cooperation (e.g., mutual visa freedom), to create a relatively peaceful international environment (as in the post-War period), to bring multi- or transnational service chains to the public (e.g., banking, insurance, credit cards, hotels, and fast food). Among other things, the *mediatization* of other peoples or places (differences), as an integral component of globalization, is particularly significant in triggering off cultural curiosity about difference.

As noted earlier, tourism involves what Aristotle describes as the "desire to see", or what Berlyne (1960) calls "perceptive curiosity". Both refer to an impulse to see or explore something curious. However, before such an impulse can be activated the individual must learn of this thing from a secondary source. Otherwise it can not become the object of curiosity. Curiosity necessarily entails a *second-hand* knowledge of something that is curious, and it is this that provokes the desire to see or explore the curious thing in person.

This second-hand knowledge of a place, a people, or a culture can be called its *image* (a more detailed discussion of image will be presented in Chapter 7). Strictly speaking, an image is different from an impression. An impression of a thing is obtained from personal visual encounters with it, whereas an image is obtained in the absence of encounters. If an impression is gained from seeing or exploring, then an image is a result of imagination before seeing or exploring, based on secondary knowledge. The elements that help form an image are hearsay, photographs, pictures, films, television, and associated imagination. Imagery is important to tourism because a favourable image may help trigger off cultural curiosity, and hence the action of presence-seeking ("be there").

Gunn (1988) has explored the image-formation process in his model of the seven phases of the travel experience: the accumulation of mental images about vacation experiences, modification of those images by further information, the decision to take a vacation trip, travel to the destination, participation at the destination, the return home, and modification of the images based on the vacation experience. What needs to be added is that returning tourists may reveal their modified images of the destinations to their friends, colleagues, and relatives, which in turn can shape those people's images of the same destinations.

According to Gunn (1988), the image formed in phase one is an "organic image", and that formed in phase two an "induced image". An organic image is formed on the basis of information assimilated from non-touristic, non-commercial sources, such as the media (news reports, magazines, books, films, etc.), education (school courses and textbooks), and word-of-mouth. Phase one is significant because, in many cases, only when people have formed a favorable image of a destination will they move on to phase two, that is, obtain further detailed information from specific commercial sources (commercial images), such as travel brochures, travel agents, and guidebooks. What is implied here is that, before tourists take an actual trip, they are already involved in a macro-process of the socio-cultural production of images about other peoples and places (differences), which constitutes an important source of personal images of these objects. At first glance tourists' decisions about where to go seem to be a matter of personal freedom and choice. In reality, however, these decisions are not entirely their own. They are, in one way or another, implicitly influenced by the *collective definition* and *representation* of other peoples, cultures, and places, of differences existing elsewhere. Of course, personal images of differences are also partly determined by people's level of education, which transfers a societal stock of images to them. The pattern of tourist flow partly mirrors the mode of cultural production of images about other peoples, cultures, and places. Among others, mediatization is one of most important mechanisms in which such images of difference are collectively defined and represented.

Therefore, the mediatization of cultures, as an integral aspect of globalization, plays an important role in the production of those images so necessary to the formation of cultural curiosity and international tourism. In early modernity the formation of images was based mainly upon the discursive or narrative mediatization of other places, cultures, and peoples, including travel writing, journals, and newspapers. Late modernity, in contrast, involves iconic and electronic mediatization, which play a more significant role in the formation of images and supply potential tourists with "media-based familiarity" with certain destinations (Greenblat and Gagnon 1983). For example, "Paris will seem far more familiar than Madrid because more movies and television shows have used it as their story location" (1983:101). Thus, international tourists may have a more favorable image of, and hence greater cultural curiosity about, Paris than Madrid. As a result, Paris will be more frequently visited than Madrid. Although the relationship between images and cultural curiosity is yet to be studied thoroughly, it is evident that there is a close connection between them. Triggered off by mediatized images, cultural curiosity tends to lead to the action of presence-seeking or exploration, namely tourist activity.

The role of the mediatization of culture in the formation of images of other places and peoples is also illustrated in "media-dependency for images". As Morley and Robins write, "We are all largely dependent on the media for our images of non-local people, places and events, and the further the 'event' from our own direct experience, the more we depend on media images for the totality of our knowledge" (1995:133). Indeed, if in a rural village most of an individual's knowledge of other villagers is obtained by personal experience, then in the "global village" knowledge of other "global villagers" mainly comes from the media, from second-hand, non-immediate sources of knowledge. In this sense, everyone is media-dependent. Personal experiences of other places, cultures, and peoples in tourism seem to be a confirmation of media images. Cultural curiosity about novelty and difference is, to a large extent, sparked by the mass media. Moreover, in response to people's demands for images of holiday resorts, these places have increasingly been mediatized. For example, on British television there are numerous popular tourism programmes, such as "Holiday" (BBC), and "Wish you were here" (ITV).

Exoticism and Modernity

At first sight, international tourism entails positive images. However, images can be good or bad, and hence not everything novel is necessarily the object of cultural curiosity. Taking "primitive" tribes as an example, why should they become the object of cultural curiosity, and why should people have "good" images of them? Under Western modernity a distinctive category is used to define cultural curiosity or "good" images of Others and differences, i.e., exoticism. As Turner and Ash put it, tourism is an

> escape from uniformity and complexity in search of the exotic and the simple. The pursuit of the exotic is directed towards other cultures (distant in time or space). The pursuit of the simple is directed towards other cultures in so far as they are seen to be more *primitive* than the home culture, but in its most common form it is anti-cultural (1975:130).

Or in Harkin's terms, "Tourism is one of several modes (anthropology is another) of discourse on the exotic" (1995:656). Bruner argues that

> The more modern the locals become the less interest they have for the occidental tourist. Tourists come from the outside to see the exotic.... Tourism thrives on difference; why should the tourist travel thousands of miles and spend thousands of dollars to view a Third World culture essentially similar to their own? (1995:224).

However, the exotic is not something simply out there. Rather it is socioculturally constructed. Although exoticism was referred to as early as Homer, in the *Iliad* (Todorov 1993:265), it is largely a cultural phenomenon associated with modernity. Pucci (1990) has traced the evolution of the term. In 18th-century France exoticism did not have all the meanings it has today. As an adjective exotic was, throughout this period, used in reference only to foreign, alien flowers and plant life, or rare objects and commodities taken from strange lands. In the 19th-century exoticism related to a sentiment or sensation attached to the individual who beheld objects of foreign origin. Thus, during this period, rather than referring to a particular object of foreign provenance, the term was confined to signifying a Western mentality and an atmosphere experienced by Western subjectivity (Pucci 1990:146).

According to the *Oxford English Dictionary,* exoticism denotes the tendency to adopt what is exotic or foreign, and the exotic character; an instance of this, anything exotic. Thus, in everyday usage exoticism can be either a mentality (or orientation), or an exotic character of things. In this chapter, the term is used in the first sense, defined as a mentality and sentiment, stemming from beholding foreign, particularly remote, cultures, forms of life, peoples, or landscapes. Applied to travel, exoticism embodies a love of and interest in difference, sensation, and novelty in other places, peoples and cultures. As Boniface and Fowler put it, "The Western attraction to the exoticism of the East has been fundamental to the formulation of many tourism packages, commencing with those to Egypt of Thomas Cook" (1993:39).

The emergence of exoticism in Western modernity is mirrored in the changing relation between curiosity and fear in regard to Others and difference. In this sense, the rise of exoticism is an indicator of the changing relationship between Western subjectivity and non-Western Otherness. Self-evidently, too much strangeness, a lack of national self-confidence, or an inability to deal with difference will cause anxiety or fear, rather than curiosity and fascination. Therefore—in a world dominated by uncertainty, chaos, threat and danger—anxiety and fear, rather than curiosity, are the dominant responses to Others and Otherness. Accordingly, foreign or alien peoples are most likely to be regarded as a threat rather than as objects of curiosity. In such circumstances, curiosity about the exotic is unlikely to occur. Once the foreign or alien people are considered as a threat, then the corresponding social–spatial strategy is either aggression and crusade, i.e., to wipe out "enemies", or withdrawal and avoidance. Therefore, in premodern societies characterized by violence and the threat of war, exoticism found no locus. With the advent of Western modernity, however, Western societies experienced domestic pacification (Elias 1982), and in association with this considered themselves "superior" to "uncivilized" or "barbarous" cultures. Based on increasingly strong national self-confidence, the term exoticism came to refer to a particular Western mentality and sentiment. Thus, roughly simultaneously with the progress of modernity, exoticism got underway in the Enlightenment and grew most noticeably in 19th century Europe (Rousseau and Porter 1990b).

Paradoxically, exoticism not only entails a stronger national self-confidence, it also indicates "an act of self-criticism" and "the formulation of an ideal" (Todorov 1993:264). It not only implies the enhanced national self-confidence associated with the achievements of modernity, but also reflects the discontents of modernity. It is based on the criticism of the dissatisfying aspects of modernity and a desire for a remote "utopian" place, which is constructed as "exotic" and "better", such as "Shangri-la", the utopian paradise in the high Himalayas described by James Hilton in his novel *Lost Horizon* (first published in 1933). Such an exotic world is imagined and portrayed in terms of the features of the home society. As Todorov observes, "The classical descriptions of the Golden Age, and, as it were, the Golden Lands, are thus derived chiefly by inverting features observed at home" (1993:265). In this sense, descriptions of the exotic are never objective; rather they are projections of Western "escape attempts", nostalgia, and ideology. As Todorov claims:

> In practice, the exotic preference is almost always accompanied by an attraction for certain contents at the expense of others.... These contents are customarily chosen along an axis that opposes simplicity to complexity, nature to art, origins to progress, savagery to sociality, spontaneity to enlightenment.... By and large, up to the end of the eighteenth century, Western European authors considered themselves to be the bearers of a culture that was more complex and more artificial than any other; they valorized others only as incarnations of the opposite pole (1993:266).

Exoticism is thus an idealization of "Others" and "savages" in terms of the loss of authenticity, simplicity, and innocence in the home society. Exoticism is nostalgic in character and that is why Western exoticism is always accompanied by primitivism (Todorov 1993:266). Western people have an interest in the exotic or the "primitive", not because they want to relive a primitive life, but rather because the exotic and the primitive constitute a frame of reference, a "live fossil" of authenticity, in contrast to which the structural constraints and complexity of their own society become transparent. The qualities of the exotic and the primitive are idealized as the attributes of paradise. The practice of idealizing of "savages" and the primitive dates back to the Renaissance (Todorov 1993).

However, at different stages, exoticism has been accompanied by quite different emotional approaches to Others and difference. Netton (1990) has distinguished two different paradigms of exoticism: the "Enlightenment paradigm" and the "romantic paradigm". The Enlightenment paradigm includes four principal elements: a dislike of aliens, a fear of threat, a fascination with the exotic, and occasionally a degree of sympathy towards its subject (1990:39). Exoticism in the Enlightenment was thus contradictory, ambiguous, and ambivalent. The Enlightenment thinkers believed in universalism, a doctrine that regards Western civilization as the embodiment of reason and rationality, and as the universal civilization that can and should be extended to the rest of the world. Thus, unassimilable and stubborn differences are often seen as danger and threat, and viewed as disgusting and distasteful. Under the Enlightenment paradigm, fascination with the exotic was often accompanied by fear and dislike of the alien and Others. Rousseau and Porter have described this ambivalent exoticism in the age of the Enlightenment as follows:

> The exotic certainly includes a bestiary of terror. It is populated with long-tailed men, clutching their heads under their arms, ichthyan-tailed mermaids, anthropophagi, sciopods, amazons, astomi (those who live from smelling apples), and all those other fabulous beasts. . .
> Yet the exotic is not solely a gallery of "landscapes of fear". It is inviting no less than threatening, exemplified by the fantasy world of Eden, the paradise garden, the Elysian Fields, Eldorado, the Happy Valley, Xanadu, the buried cities of the Incas, the lost tribes, the idea of the ultimate aborigine, the noble savage, the state of nature; it is the Africa from which there is always something new, or that *terra incognita* utterly unsullied by man, the virgin whiteness of Antarctica (1990b:4–5).

Exoticism therefore includes an element of fantasy. It is a socioculturally produced utopia of difference.

With the progress of modernization, discontent with the dark side of modernity became increasingly evident. Correspondingly, what Netton (1990:39) calls the "Romantic paradigm" of exoticism began in the nineteenth century. The connotation of the term "romanticism" is thus extended from a love of nature to a love of uniqueness and diversity (Rousseau and Porter 1990b). Thus, romanticism became not only a demonstration against modernity's (industrialization's) "violence" towards nature (see chapter 3), but also a movement against Western modernity's universalization and associated assimilation of cultural differences. According to Rousseau and Porter the romantic movement was a revolt against the "uniformitarianism" of Western modernity:

> Today's cultural leaders, academic and popular, are no longer so enthused by such "uniformitarianism". Ever since the Romantic revolt, the folly and bad faith implied by the "tyranny" of universal standards—the authority of the Ancients and of Classical rules, the despotism of polite French taste, and so forth—have been objects of reproach. Romanticism proclaimed the sovereignty of uniqueness and holiness of diversity. And today, it is "difference" that dominates our discourse (1990b: 2).

If in the Enlightenment paradigm the function of self-criticism is not very clear, in the romantic paradigm of exoticism this function is obvious enough. Romantic exoticism is an idealization of primitivism and the exotic as well as a cultural criticism of the negative side of modernity. Moreover, in the age of post-colonialism, romantic exoticism may also contain an element of what MacCannell (1992:20) calls "modernity's guilt", a guilt of, say, the Western white's earlier maltreatment of the so-called "primitive" American Indians. International tourism in the Third World today, may be an unconscious Western drive to alleviate this guilt. It is, so to speak, a ritual respect for difference. However, the ritual respect for Others still hides some distortion about them, and international tourism may itself paradoxically threaten the survival of the authenticity of Others (MacCannell 1976, 1992).

Exoticism and Ideology

Exoticism, as a love of difference, uniqueness, or diversity, implies questioning the taken-for-granted order at home. However, exoticism is itself contradictory. It is both a self-criticism and the idealization of difference and Otherness, and it conceals ideological biases and cultural stereotypes about Others and differences.

To risk oversimplicity, there exist, ideal-typically speaking, three kinds of exoticism. The first, is *"neutral exoticism"*, refers to a "pure" romantic love of difference, uniqueness, and diversity between equals. The second, *"bottom-up exoticism"*, refers to weak nations, adoration of and curiosity about advanced nations, whose skyscrapers and electronics in metropolitan centres such as New York, Hong Kong, or Tokyo are seen as exotic (Todorov 1993:266). This kind of exoticism may possibly contain a self-abasing element. The third, *"top-down exoticism"*, refers to advanced countries' interest and curiosity about "primitive" or "exotic" cultures, ethnic groups or nations, and indigenous forms of life. Most of Western exoticism belongs to this last type. Such "top-down exoticism" implies a sense of *self-superiority*, a psychic position in which one stands high to look down at something exotic, alien, and strange. As Turner and Ash point out, "The natives, of course, would not be so picturesque if they were prosperous and educated" (1975:201). Therefore, "top-down exoticism" hides an implicit preassumption: "we the civilized" people are superior to "them the exotic". Love of the exotic might not be directly generated by the consciousness of self-superiority, it is rather generated by the consciousness of disappointment with some aspects of one's "superior" civilization. However, the consciousness of self-superiority always lurks behind the mentality of exoticism. And this self-superiority embodies the colonial impulse. As Boniface and Fowler reveal:

> Tourism feeds on the colonial impulse. Part of the appeal, the *frisson*, of traveling to strange lands is the opportunity that it may afford to patronize the poor native unfortunates who may know no better way of life than that of their homeland. Tourism, in many ways, is a sort of neo-colonialism (1993:19).

Said (1978) has revealed that Western discourses on the Oriental conceal colonialist and imperialist prejudices against stereotypes of the Other. The same is true of the Western representation of exotic peoples.

Exoticism involves two contradictory definitions of Others and difference in terms of the projection of Western consciousness. When the disappointing aspects of Western modernity come to the fore, Others and difference are *idealized* as "noble savages", and exotic places are defined as paradises to which one can imaginatively escape. This mentality is exhibited as exoticism. By contrast, when the triumphant aspects of Western modernity come to the fore, Others and difference are *reproached* as barbarous and backward, and exotic places are then regarded as alien places of horror and danger. This orientation is embodied in colonialist ideology. These two mentalities are often interlinked, and it can be difficult to separate one from the other. No matter which way Others and difference are defined, the image of them is distorted. In the case of exoticism, some elements, for example erotic freedom, are idealized at the expense of others, for example, their daily drudgery and poverty. In the case of colonialist ideology, some factors, such as cruelty, are exaggerated at the expense of others, such as rationality. One of the pitfalls of the colonialist distortion of exotic locals is the justification of colonial rule. As Kabbani points out:

> In the European narration of the Orient, there was a deliberate stress on those qualities that made the East different from the West, exiled it into an irretrievable state of "otherness". Among the many themes that emerge from the European narration of the Other, two appear most striking. The first is the insistent claim that the East was a place of lascivious sensuality, and the second that it was a realm characterized by inherent violence ... it was in the nineteenth century that they [these themes] found their most deliberate expression, since that period saw a new confrontation between West and East—an imperial confrontation. If it could be suggested that Eastern peoples were slothful, preoccupied with sex, violent, and incapable of self-government, then the imperialist would feel himself justified in stepping in and ruling. Political domination and economic exploitation needed the cosmetic cant of *mission civilisatrice* to seem fully commendatory (1986:5–6).

The Enlightenment paradigm of exoticism was obviously intertwined with such colonialist and imperialist prejudice against Others (Rousseau and Porter 1990a). This situation was partly a result of the world-view of Enlightenment that persisted in the universalization of Western rational order. Such prejudices are less obvious in the romantic paradigm of exoticism. However, after the old, crude, colonialist prejudices against Others were driven out, a neocolonialist bias was invited in. While old colonialism as exhibited in the Enlightenment paradigm of exoticism, explicitly expresses fear and disgust of Others, the neocolonialism implicit in the romantic paradigm of exoticism, on the contrary, stubbornly insists on the absolute *preservation* of the exotic, and hence the "museumization" of "primitive" peoples. Any sign of change, particularly modernization, on the part of "primitive" or "exotic" peoples is seen as destroying local cultures and is hence unacceptable. What is implied here is that exotic peoples should remain the same, "primitive" and "backward", *because "I" need them to be so*. If these peoples modernize their lives, then where will "I" find the exotic to satisfy "my" romantic curiosity? Such a neocolonialist bias is also exhibited in the discourse on authenticity in tourism (see *chapter* 1).

Where modernization has changed local cultures and resulted in "staged authenticity", because authenticity of the original is no longer possible, this has offended and upset many Western tourists and academics. Their "imperialist nostalgia" (Rosaldo 1989) is exhibited as mourning for what has been destroyed, that is, the disappearance of primitive authenticity as a result of modernization. For them, if possible, authentic and exotic people must be "frozen" in time or preserved in segregated and insulated zones in order for them to remain authentic and exotic, and therefore still backward and poor. In short, for the sake of romantic exoticism they hope that exotic people will have no chance at all of access to the "monster" of modernity, which has spoiled so many authentic cultures. This does not imply that cultural traditions should not be preserved. Rather, the point at issue is that the preservation of tradition should not be achieved at the expense of modernization, or against the natives' will to improve their lives. In reality, once modernization is set in motion the preservation of tradition is always relative, because realistically nothing in the world can remain unchanged.

Clearly, in relation to such ideological biases, the representations that Western travelers produce about Others or the exotic are often different from the perceptions that locals have about themselves. What travelers find interesting and exotic may be seen by locals as far from true representations of themselves. In this sense, the narrative undertaken by Western travelers, for example, may impose "image violence", misrepresentation, upon "exotic" locals. As far as the larger mechanisms of image production are concerned, the flow of distorted images about Others mirrors the asymmetrical power relations of international communications and representations, between weak nations and advanced nations. About this, Morley and Robins write:

> given the largely one-way nature of the flow of international communications, it is the Western media which arrogate to themselves the right to represent all non-Western Others, and thus to provide "us" with the definitions by which "we" distinguish ourselves from "them" (1995:134).

Armed with such images of exotic peoples and places, international tourism to Third World destinations is often embodied as the *projection* of Western consciousness and utopia rather than exploration into the reality of people's existence. Exoticism in international tourism is therefore only an interest in filtered and selected signs of symbolic Others and differences.

International Tourism and Spaces of Difference

For international tourism to occur, two of a number of conditions need to be satisfied. First, a national subject who has identified with a nation must be able to find support (economic, intellectual, spiritual, political or even military) from the nation-state when abroad, particularly in remote destinations (Kabbani 1986:1). Second, there needs to be peace in international relations—no hostility, danger, or threat of war between heterogeneous nations. As previously seen, both nationalization and globalization, as two kinds of social spatialization (social–spatial strategies) of modernity, help to satisfy these two conditions. Nationalization creates a space of consensus and citizenship, and the culturally confident national subjects (Alter 1985). These national subjects, sustained by empire or nation-state (structural condition) and national self-confidence (cultural condition), show cultural curiosity towards the outside world. In comparison globalization, particularly that of late modernity, tends to remove and reduce hostility and threats between heterogeneous nations and national cultures, based on enhancing interdependency and cooperation between nations; hence the removal of spatial barriers creates accessibility to other countries. The interplay of these two social–spatial strategies constitutes one of the preconditions for international tourism. International tourism thus thrives on spaces of difference, spaces differing from both the "space of *consensus*" and the "space of *hostility*".

International tourism allows tourists voluntarily to enter geographically, socially, and culturally "differential spaces" (Lefebvre 1991), namely spaces untied from the social controls and norms that exist in the spaces of the home society. Such spaces are unusual, unfamiliar, and novel, and are therefore appealing, thrilling, and exciting. Kabbani (1986:21) characterizes the difference between the space of the West and that of the East: "the West is social stability, the East pleasure". However, as previously noted, unfamiliarity and uncertainty must to a certain extent be under control, otherwise they will cause fear and anxiety. But too much familiarity and certainty is not conducive to pleasure and excitement. Therefore, there must be a balance between familiarity and unfamiliarity, certainty and uncertainty, similarity and difference. Differential spaces are therefore always relative. These spaces can be further characterized as "spaces of difference for sightseeing" and "spaces of difference for strangers", respectively .

Spaces of Difference and Sightseeing

International tourism involves searching for the exotic that typifies difference. Thus, international tourism offers access to spaces of difference, which are in contrast to the spaces of home. Whereas rational order dominates the spaces of home, it is pleasure, particularly sensational pleasure, that characterizes spaces of difference. International tourism also reflects modern people's "will for difference". They want to make a difference to their daily routines and everyday environments. Thus, they travel and enter the spaces of difference. A visit to the exotic is an experience of difference which constitutes a temporary transcendence of the constraints and malaise of the home space.

The exotic is a source of sensation. Part of the reason for this is that the exotic is often defined as the erotic (Porter 1990). More generally, the exotic exhibits qualities that are alien, strange, and novel, and hence lead to pleasurable, exciting, and stimulating experiences. Hazilitt described such experiences in 1821:

> There is undoubtedly a sensation in travelling into foreign parts that is to be had nowhere else: but it is more pleasing at the time than lasting. It is too remote from our habitual associations to be a common topic of discourse or reference, and like a dream or another state of existence, does not piece into our daily modes of life (quoted in Kabbani 1986:87).

The sensation of the exotic is also to be found in foreign physical environments. For example, for British tourists the desert is exotic and sensational in comparison with the green, grassy environment of their own habitats. Tuan explains the reason for this:

> the desert, despite its bareness, has had its nonnative admirers. Englishmen, in particular, have loved the desert. In the eighteenth and nineteenth centuries, they roamed adventurously in North Africa and the Middle East, and wrote accounts with enthusiasm and literary flair which have given the desert a glamour that endures into our time, as the continuing interest in the exploits of T. E. Lawrence shows. Why this attraction for Englishmen? The answers are no doubt complex, but I wish to suggest a psychogeographical factor—the appeal of the opposite. The mist and overpowering greenness of England seem to have created a thirst in some individuals to seek their opposite in desert climate and landscape (1974:xii).

The sensation of the exotic is, however, more often derived from exotic customs, cultures, and peoples, particularly when they are remote in both time and space. For example, in 1503 Amerigo Vespucci described his encounter with South American Indians:

> They have no cloth, either of wool, flax, or cotton, because they have no need of it; nor have they any private property, everything being in common. They live amongst themselves without a king or ruler, each man being his own master, and having as many wives as they please. The children cohabit with the mothers, the brothers with the sisters, the male cousins with the female, and each one with the first he meets. They have no temples and no laws, nor are they idolaters. What more can I say! They live according to nature (quoted in Todorov 1993:267).

For most Europeans at that time the sensation of the exotic was second-hand. It was based on hearsay; the certainty of its contents was beyond the reader's control. For the exotic to become really sensational it must therefore be physically and geographically accessible. In other words, it must be seen, or confirmed, in person. The exotic becomes sensational only when it is the object of sightseeing and people yearn for such sightseeing. Seeing is a certain way of control. If something has not been seen, it is not yet under control. Therefore, to control (or confirm) the exotic, the exotic must first of all be seen. Hence, the origin of sightseeing has to do with the drive to explore and control unknown situations. In association with this drive, a modern epistemology develops which emphasizes the eye as the primary sensual organ in the collection of data upon which valid knowledge is based (Adler 1989a; Heidegger 1962:215–216). It can be argued that sightseeing is essentially modern, corresponding to the generalized intellectual ethos of modernity. According to Adler (1989a), the history of travel in Europe underwent a change from travel as discourse, which was systematically practiced in the early sixteenth century, to travel as sightseeing from the sixteenth century onwards. "With inevitable juridical reference, travel is praised through a favourable contrast between 'eyewitness' and 'hearsay' as legally admissible evidence and ground for valid judgement" (1989a:11). The advent of sightseeing was roughly simultaneous with the advent of modernity, particularly with the emergence of Baconian and Lockeian philosophical epistemological subjectivity. Baconian and Lockeian epistemology and modern sightseeing share a common feature; both give "preeminence to the 'eye' and to silent 'observation'", both insisted on the "ascendancy of the eye over the ear" (1989a:11).

Sightseeing is not a kind of "cool" scientific seeing. On the contrary, sightseeing, or what Urry (1990a) calls the "tourist gaze", is a kind of "hot" seeing (Uzzell 1989). It is therefore quite natural for anthropologists to link sightseeing to rituals (Graburn 1989). Sightseeing is a ritual exploration where sightseers explore the space of difference with an acute eye, an eye that can discover what is taken for granted by locals and hence those things of which locals are unaware. This difference between natives' and visitors' seeing is noted by Tuan:

> Generally speaking, we may say that only the visitor (and particularly the tourist) has a viewpoint; his perception is often a matter of using his eyes to compose pictures. The native, by contrast, has a complex attitude derived from his immersion in the totality of his environment. The visitor's viewpoint, being simple, is easily stated. Confrontation with novelty may also prompt him to express himself. The complex attitude of the native, on the other hand, can be expressed by him only with difficulty (1974:63).

In the contemporary age sightseeing takes on new characteristics, that is, it is linked to the camera with which sightseers take photographs of difference (Chalfen 1979; Crawshaw and Urry 1997; Harkin 1995; Sontag 1979; Urry 1990a:ch.7, 1992:4–5). Horne claims that "the camera and tourism are two of the uniquely modern ways of defining reality" (1984:121). Indeed, the camera is an almost indispensable tool in the ritual of sightseeing. The camera helps embody the ritual features of sightseeing by catching the moments of what has been seen. Not only is it an extension of sightseeing, but it also equips the sightseer with more sensitivity, or more of an artistic and aesthetic orientation towards tourist sights (Crawshaw and Urry 1997). More importantly the camera objecti-fies memories of sightseeing. The camera enables after-effects of sight-seeing by recording the history of itineraries and experiences, and then reinforcing the sense of belonging when shared emotion is felt while viewing the family album together.

However, sightseeing, as a curious gaze, may sometimes have proble-matic social consequences. Once the object of sightseeing is people, rather than things, a tension arises between "seeing" and "being seen", or between "touring" and "being toured". Sightseers' seeing may be socially offensive, if not physically constraining, because it may be a "zoologizing" kind of seeing that treats those viewed as curious animals in a zoo, or as "pan-zoos". This "zoologizing" seeing can be thought of as an embodiment of "cannibalism" in the contemporary world (MacCannell 1992), or "visual rape".

Spaces of Difference and Strangers

From the perspective of natives, tourists, who enter the space of differ-ence are similar to Simmel's "strangers" (Böröcz 1996:5–9; Dann 1996a:12–15; Greenblat and Gagnon 1983; Kaplan 1960:216; Nash 1989:44; Schmidt 1979:460; Watson and Kopachevsky 1994:653). Greenblat and Gagnon claim that "Travelers are temporary strangers—strangers who have chosen to enter geographically, personally, socially, and culturally unfamiliar territory" (1983:91). Travelers are also volun-tary strangers who "move into worlds marked by different degrees of strangeness and novelty. This sense of differences is a major attraction of travel—the break of the routine, the regular, and the expected" (1983:106). Tourists' experiences of "strangerhood" are like the more general experience of modern urban dwellers who must regularly deal with strangers (Greenbalt and Gagnon 1983:91; Machlis and Burch 1983:669).

The "stranger" is a sociological category put forward by Simmel to describe a kind of social relation. Of strangers, Simmel writes:

The unity of nearness and remoteness involved in every human relation is organized, in the phenomenon of the stranger, in a way which may be most briefly formulated by saying that in the relationship to him, distance means that he, who is close by, is far, and strangeness means that he, who is also far, is actually near. For, to be a stranger is naturally a very positive relation; it is a specific form of interaction (1950a:402).

Strangers are thus physically close yet spiritually distant (Bauman 1990b:66). To put it another way, they "are, as it were, neither close nor distant. Neither a part of 'us' nor a part of 'them'. Neither friends nor foes. For this reason, they cause confusion and anxiety" (1990b:55). These characteristics of strangers are also applicable to tourists. As Schmidt points out, "As a result of both integration and insulation, the tourist, similar to Simmel's stranger, is both near and far from the new environment" (1979:460). The tourist is physically close to but socially distanced from the native community. "He does not have to become a member of that culture and be accepted by the group. He has not abandoned his former social position, and the tour will guarantee that he won't" (Schmidt 1979:460).

For locals, tourists come without invitation and are beyond their control. Therefore, touristic comings and goings are difficult for locals to manage and control. From the perspective of tourists, however, locals, as well as other tourists, are strangers too. "All travellers are presenting themselves before audiences whose responses are relatively unpredictable and whose potential for misunderstanding is extremely high" (Greenblat and Gagnon 1983:97). Therefore, the experience of "strangerhood" in tourism raises the problem of management of novely and unfamiliarity in order to create order out of it. In Greenblat and Gagnon's terms, "As the traveler seeks out the different, techniques are required to reduced the unfamiliar, to make what is strange somehow understandable, if only through successful management" (1983:99). Greenblat and Gagnon classify three types of management: resource management, management of social relations, and identity management. For tourists, a visit to the space of difference means a deroutinization of social relations, and confrontation with unfamiliar and uncertain social situations, social relations and environments, which may cause certain crises in their lives. However, part of the pleasure relating to this situation is "the triumph over or successful management of these self-induced crises or problems" (1983:96).

Therefore, for "differences" to become a source of pleasure, a certain degree of control over them is necessary. As varying degrees of novelty and uncertainty in tourism are linked to corresponding degrees of effort in tackling this problem, tourists can be classified according to the type of tourism chosen in relation to the degree of certainty. Cohen (1972) has classified tourism into institutionalized and non-institutionalized tourism. Both the organized tourist and the individual mass tourist choose institutionalized forms of tourism, while non-institutionalized forms of tourism are chosen by the explorer and the drifter. These different forms involve different degrees of control over novel situations by tourism brokers, and they require different degrees of effort in the management of novel experiences by tourists themselves. For mass tourists the guided tour is often the preferred choice. As a form of management of novelty and strangeness, the guided tour has three characteristics: psychological security is provided; economic simplification and security are facilitated; and interaction with natives is minimized (Schmidt 1979:443). Thus, for mass tourists the search for difference is conditional on the successful management and control of novelty, strangeness, and uncertainty arising in the space of difference. Tourists are not homogeneous, but they all require an optimal balance between strangeness and familiarity, and stimulation and tranquility (Yiannakis and Gibson 1992).

With the globalization and commodification of tourism the space of difference is more and more purged, controlled, and insulated. As Bauman notes:

> the strange is tame, domesticated, and no more frightens; shocks come in a package deal with safety. This makes the world ... obedient to the tourists' wishes and whims ... what the tourists buys, what s/he pays for, what s/he demands to be delivered ... is precisely the right not to be bothered, freedom from any but the aesthetic spacing (quoted in Jokinen and Veijola 1997:33).

As a result tourism can no longer be defined as "exploration" in its original sense, not to mention "adventure". It has, ironically, become what Schmidt (1979) calls "insulated adventure".

As the management of "strangerhood" tourism is a superficial and transitory form of interaction between tourists and locals. Correspondingly, tourists' search for difference and the exotic remains at the level of sensation and the sense of the picturesque, far from the level of mutual understanding. Kaplan distinguishes two kinds of American traveler: comparative strangers and emphatic natives. Whereas the latter seek, as best they can, to put themselves in the place of those whom they visit, the former

> never, or seldom, leave their own familiar ideas and judgments. They find security wherever they may be in what is called ethnocentrism, the application of one's own standards to other situations: their own are always superior to those of others. They view, but do not understand (1960:216).

The issue at stake is that, in modernity, differences, the exotic, and novelty are presented as commodities for mass consumption. They are the consumer goods of the sensational and picturesque, rather than objects for understanding. Turner and Ash have characterized tourists this way:

> They are paying money in order to be waited upon; the vast majority do not really want to get know their hosts, and when they do, they are hampered by the genuine problem that there is a limit to how much one can discover in a fortnight about someone speaking a different language. Their perception of local realities will therefore be superficial and prejudiced, just as the locals in return will view the tourist with suspicion (1975:245).

Although the consequences of tourist–host contact vary according to numerous factors, there are indeed some interaction difficulties in cross-cultural tourist–host communication. These difficulties can often give rise to strain, a sense of loss, a feeling of being incompetent, rejected, or frustrated, a confusion of values, anxiety, an increase in irritation or criticism—emotions which can be termed "cultural shock" (Reisinger 1994). Thus, the pleasures derived from the visual consumption of difference and the exotic are conditional upon a certain degree of social and personal *control* over the differences. As mentioned above, package tours have emerged as the form of tourism that is most often used in order to avoid cultural shock in spaces of difference, a situation which in turn limits advances in cross-cultural understanding.

In summary, international tourism assumes that behind the tourist stands a confident nation, in front of the tourist exists a safer, more cooperative, and peaceful world, and between tourists and their destinations lie the various brokers who arrange itineraries for them, and hence reduce the potential anxiety and fear of overseas travel.

Tourism involves traveling to other places, it involves motivating cultural curiosity about differences and the exotic. Both curiosity and exoticism are the outcome of sociocultural construction, which is closely bound up with the social–spatial strategies (or social spatialization) of modernity; nationalization on the one hand and globalization on the other. Nationalization creates a potential space for cultural curiosity about Otherness on the grounds of homogenization of the space of national identity and the removal of fear of Others. Turning this potential space into realistic space requires a peaceful global context. Globalization in late modernity, then, produces accessible, hostility-free, or more "friendly" international contexts which favor international tourism or sightseeing. In so doing, globalization either leads to the "villagization" of the world, which integrates more and more remote places into the tourist reach, or promotes the "mediatization of culture" on a global scale, which helps trigger off cultural curiosity about the exotic. Therefore, in terms of social–spatial strategies (social spatialization), modernity gives rise both to a curious subject who is ready to move about and travel to other places and countries, and to a new context which enables the tourist to travel abroad without fear of threat.

A specific category to express cultural curiosity about Others and difference is exoticism. Exoticism is value-laden and subject to ideological biases. Its emergence as an interest in the "primitive", or the Oriental has to do with the social–spatial condition of modernity. It implies both a criticism of the constraints in the space of the home society and an idealization of the space of difference. Such an idealization may be a distortion of reality, however.

International tourism offers people access to the space of difference for sightseeing, and hence satisfaction of their cultural curiosity. As "differences" are experienced as a source of sensation and pleasure, they thus become consumer goods for mass consumption rather than objects for real understanding. Furthermore, the space of difference involves a world of strangers, but many daily life skills are not geared to cope with novel situations; thus, international tourism raises the problem of how to manage novel experiences. As a result, the degrees of novelty, uncertainty, and strangeness relating to the space of difference are reduced by different forms of tourism, including all-inclusive package tours. International tourism, when taken in the latter form becomes merely an "insulated exploration" or an "insulated adventure".

PART 3

MODERNITY, TOURISM AND SEDUCTION

Part Three deals with the conditions of modernity that lure tourists towards leisure and pleasure travel. The sociogenesis of tourism involves not only the "push" of modernity, as discussed previously, but also its "pull". The "pull" of modernity includes structural factors, such as ever improving transportation technology and communication, marketing skills, and increased disposable income and time. However, as these factors are well documented they will not be touched upon here. Rather, Part 3 will concentrate on cultural factors that entice. These include touristic representations, such as images and discourses, and touristic consumer culture. Tourism is treated as a consumer culture because it is essentially a consumer activity, one that involves cultural orientations and values regarding touristic consumption.

Touristic consumption includes the consumption of representations (images and discourses), experiences, and the sign values of products. Both images and discourses help bring about imaginings and anticipations of destinations, which are part of the pleasure of tourism (Urry 1990a). Touristic experiences are consumed in the ongoing journey; they constitute the essence of tourist pleasure—the consumption of "a good time". Touristic consumption is also of symbolic significance. It contributes to status differentiation and functions as a symbolic demarcation of social locations. Tourism is thus pursued partly because of its semiotic function of defining and communicating social identity. In this sense, touristic consumption is a norm which puts social pressure on, even stigmatizes, non-consumers. In sum, touristic consumption is a cultural construction.

Four chapters are devoted to the pull of modernity. In Chapter 7, images are discussed and in Chapter 8 discourses are analyzed. Both chapters take a sociological approach to the issue of tourist representations (images and discourses). Chapter 9 deals with the experiential dimension of touristic consumer culture, and Chapter 10 focuses on its symbolic features.

Chapter 7

The Lure of Images

As discussed in the previous chapter, tourism involves the space of difference. For tourists, to enter this space implies a pleasurable experience, that is to say, they *anticipate* that pleasure will stem from travel. At the same time, as tourism is an "experiential product", tourists' purchase of this product is based on *confidence* in its qualities. Anticipation and confidence are thus the components that constitute the tourist image.

After discussing the issue of tourist motivations, which involve the questions "why travel?" and "what kind of travel?", a more specific issue arises with regard to "where to go?" and it is this question which imagery addresses. Images are important in both choice of destination and for the destination itself, and these dimensions have caught academic attention. However, the emphasis in the literature on tourist imagery seems to be placed upon the psychological, practical, and management aspects of image formation. The wider context in which tourist images evolve is usually ignored. For this reason, this chapter discusses images and representations in tourism promotion from a wider sociological perspective. Images are examined in relation to society and culture, and particularly to the condition of modernity. This chapter consists of five sections. The first discusses the role of image packaging in modernity. The second defines the term image. The third elaborates the relationship between ways of seeing and tourist images. The fourth deals with the role that culture plays in the creation of tourist images in advertising. Finally, the fifth section explores the seductive power of images in relation to the wider context of society and culture.

Modernity and Image Packaging

"Seeing is believing." This has long been commonsense knowledge in Western society, characterizing not only legal procedures (e.g., the important role of the witness), but also, more importantly, a widespread positivist ethos that underpins various practices of science and communication. Correspondingly, there are two interrelated social practices. One is the practice of concealing private affairs or indecent behavior, such as defecating, from public view (Elias 1978). The other is the practice of publicity, promotion, and advertising by individuals and commercial or governmental organizations, which involves deliberate visualization and public presentation (Boorstin 1964; Goffman 1959). In short "bad" (negative images) must be hidden, and "good" (positive images) must be made visible. Such a strategy of presentation is mostly related to anonymous urban society governed by the mass media.

Cultural visualization and presentation thus constitute an important strategy in contemporary society (Simmel 1990). With the constant evolution of visual technology, such as photography, color printing, film, television, video, and the Internet, society has increasingly been turned into images and spectacles. Relatedly, "visual persuasion" and "visual manipulation" (Rutledge 1994) have become new sources of power. In Fyfe and Law's terms, "the making of the twentieth century world has had a distinctively visual aspect" and "modernization has involved the eye" (1988:3).

"In everyday consciousness, images become important as never before" (Boorstin 1964:200). As a way of cultural visualization, "image packaging" has become an increasingly important strategy for both individuals and organizations. Image or "impression management" is, for instance, an important component of the presentation of self in everyday life (Goffman 1959). The image, as exhibited in brand names, is also a commercial strategy of distinguishing one kind of commodity from another in order to catch the consumer's eye. Such imagery is used by corporations in setting up positive identities in commercial competition (Boorstin 1964). The image is, in short, both a means of establishing distinction as opposed to homogeneity, and a means of "competition for attention" rather than anonymity (Boorstin 1964:199).

Image packaging is one of the cultural conditions of capitalism. As a system of commodity production, capitalism entails not only commodity competition but also image competition, namely "competition for attention". It is a banal observation that the consumer images of commodities and corporations are crucial to survival under capitalist market competition; otherwise corporations will lose their market share and fail in their business. Therefore, to establish, maintain, and manage a good public image is a key to success. In other words, to be attractive a commodity requires not only functional material packaging but also image packaging through designs, brand names, slogans, and advertising. The latter helps attach what Baudrillard (1988) calls "sign values" to commodities. Geared to the system of capitalism, the labor force vies for employment through image competition as well. Work applicants do their best to impress managers in order to land a desired job. Image competition is also extended to political life. With the unprecedented penetration of the mass media into Western political life, a good media image is crucial for the success of politicians. Today's politics is to a large extent image politics. As Boorstin so clearly states in reference to American polity: "More important than what we think of the Presidential candidate is what we think of his 'public image'. We vote for him because his is the kind of public image we want to see in the White House" (1964:204).

Image packaging most often appeals to the mass media because media images are the most influential. Today, ubiquitous media images constitute a "symbolic environment" (Castells 1996), which is, so to speak, differentiated from both the physical and the social environment. It is a commonsense experience that in the symbolic environment people are bombarded by a constant flow of images. For example, in the USA in the late 1980s, television presented 3600 images per minute per channel and, according to the Nielson Report, each adult on average spent roughly 4.5 hours a day watching TV (1996:333). Daily exposure to the media cultivates a habit of "image-thinking" (Boorstin 1964:185, 187, 197, 214) in both individuals and organizations. It is also claimed that Western society has shifted from a linguistic, word-dominated culture to an iconic, image-dominated culture (Dondis 1973; Postman 1985:9). Irrespective of whether this statement is overexaggerated, clearly nobody denies the importance of images in contemporary economic, political, social, and cultural life.

Interestingly, places also vie for advantageous images in attracting both capital and tourists under the condition of accelerating globalization. The latter, namely the tourist image, is the focus of what follows. For this reason a conceptual definition of imagery in general and tourist imagery in particular is necessary at this stage.

Defining Image and Tourist Image

"Image" is a term "with vague and shifting meanings". Thus, "attempts to define the term operationally are doomed to failure since the wide and everyday use of the word swamps the scholar's attempts to use it in a specific limited manner" (Pearce 1988:162). Methodologically speaking, images are highly subjective and sometimes even subconscious; as such they are difficult to express, measure, and quantify (Telisman-Kosuta

1989:559). However, it is necessary to review briefly the usage of "image" in various contexts.

The English word "image" comes from the Latin *imago*, which in turn is derived from *imitari* ("to imitate"). Therefore, "an image is an artificial imitation or representation of the external form of any object, especially of a person" (Boorstin 1964:197). Similarly, Berger defines the image as "a sight which has been recreated or reproduced" (1972:9). Images are thus the visual representation of the appearance of something that is absent (1972:10). Karl also argues that "The image is not the thing"; it is only a representation, which cannot be confused with the material object it represents (1994:198). Fox (1994a:x) goes a step further in classifying two kinds of visual representation. First, the image is an actual and pictorial representation. Second, it refers to a mental representation without the actual stimuli being present, namely images in mind or imaging, both of which are interrelated. For Fox, "images are the integral components of thinking, speaking, listening, writing and reading"; they constitute the "primary underlying structure in language, media and mind" (1994a:x). Accordingly, a major meaning of image is the visibility or visual imaginability of objects, which can be represented either in the media (pictures, photos, screens, sculptures, etc.) or in the mind (when the objects are absent from vision).

Image is also used metaphorically or rhetorically at a more abstract (or invisible) level. For example, when used in a literary sense (from the 18th century onwards) it is defined "as a feature of rhetoric—a description or evocation of a scene or action so vivid that it makes the listener almost believe he is actually witnessing it" (Furbank 1970:26). In this sense, an image can be set up with words and language, and not necessarily by visual means (Dann 1996c). Second, it is used in the context of business to refer to concrete trademarks, brands, designs, and slogans that act as indices of certain commodities, consumer services, and corporations (Boorstin 1964). Such an image is also applied in consumer psychology to refer to "the holistic impression of the relative position of a brand among its perceived competitors" (Poiesz 1989:467). Third, it is used as a metaphor to signify "a studiously crafted personality profile" (Boorstin 1964:186) of an individual, institution, corporation, product, or service, which is worthy of, and gains the public's trust and beliefs. Finally, it is used in the sense of reification to conjure up the appearance or representation of an abstract conceptual substance with no concrete presence. Examples might include the image of God, of happiness, of the middle class, of the modern, and of a nation.

In addition to the primary meaning of image, namely, visibility either in pictorial or mental representation, the term also contains another set of meanings, including attitudes, beliefs, conceptions, and feelings that a person has about an object, which are mixed with visual perceptions or impressions of the object. In Chon's terms, "an image is the net result of the interaction of a person's beliefs, ideas, feelings, expectations and impressions about an object" (1990:4).

An image has several characteristics. First, it is vivid and concrete, with visual components. Even if something is invisible, it is still possible to turn it into an image in visual and semiotic terms. For example, a private car can conjure up the image of a modern style of life; private detached houses with gardens give an image of the middle class; smiling faces present an image of happiness; the image of roses conveys love and passion (Barthes 1972). Second, an image is social. It presupposes an audience or receivers. Quite the opposite to material objects, images can only exist in and through the minds of this audience and are therefore intersubjectively or publicly held (Pearce 1988:162). Third, an image is partial and simplified, its visual components are selective, never a full and complete representation of an object. Fourth, the selection and combination of these visual components are arbitrary and subject to preferences or customs. As a result, images are usually biased and distort reality. Fifth, images are stereotypes (Pearce 1988). Once formulated, they are relatively slow to change. Sixth, images are ambiguous (Boorstin 1964:193). Although an image is concrete, there is always a certain vagueness about it. Room is left for the imagination. Seventh, images often charged with feelings and emotions. A "bad" or a "good" image of an object may lead to quite different responses to that object. Finally, images can be planned, built and communicated, usually by means of the mass media. On the one hand, the formation of images on the part of individuals can be passive, and they are easily subject to the influence of media bias. On the other hand, certain organizations with capital or power, such as large corporations and governments, can actively take advantage of this passive image-formation process and build an advantageous image by means of available resources (e.g., advertising).

What has been said of the image in general is also applicable to tourist destinations, i.e., the tourist image. "All places have images—good, bad, and indifferent—that must be identified and either changed or exploited" (Hunt 1975:7). The role of destination imagery in tourism development is highly significant (Chon 1990; Hunt 1975). Part of the reason for this is that the images tourists have of destinations constitute "pull" factors (Gartner 1993:193). However, the term "destination image" or "place image" is also vague. A place has many different aspects: polity, economy, culture, history, physical and built environments, population, ethnicity, weather, territory, and so on. Usually, people are not able to have images of all these features, only an overall image of selective features of a place or destination. Therefore, individuals can only have a *tourist image* of a potential destination, rather than a *resident's* image of the same place.

What, then, is the tourist image of a place? Crompton (1978) defines a destination image as the aggregate sum of beliefs, ideas, impressions, and expectations that a tourist has about a destination area. Gartner and Hunt (1987) more simply define tourist images as impressions held about a destination. Barke and Harrop further distinguish place images from place identities. "Places also have identities. The 'identity' may be regarded as an objective thing; it is what the place is actually like" (1994:95). Images are strongly influenced by objective place identity. They are different from the latter, however. "Images may exist independently of the apparent facts of objective reality" (1994:95). Indeed, as visual and mental representations of a place, images are the "symbolic transformation" of a place (Ashworth and Voogd 1990). In addition, Gartner (1993) and Dann (1994a) classify and analyze three different but interrelated components of the tourist image: cognitive, affective, and conative.

Ways of Seeing and Images

In *Ways of Seeing*, Berger claims that "[t]he way we see things is affected by what we know or what we believe" (1972:8). For example, a widespread belief in hell in the Middle Ages led people to perceive the sight of fire in a way quite different from today's generation: whereas modern persons see only a chemical phenomenon, what individuals in the Middle Ages saw meant the threatening power of hell (1972:8). Indeed, as Kuhn (1962) and many other philosophers of science point out, even among scientists characterized by "cool" observation, the same phenomenon may mean quite different things. Their various ways of seeing are the result of training within different groups with different worldviews, or what Kuhn calls "paradigms". Foucault (1976) also analyzes how the medical gaze is socially constructed as a particular way of seeing, which is supported by certain institutions such as a clinic. In brief, ways of seeing are affected by worldviews, values, and beliefs that are the result of socialization and also of mass communication.

Ways of seeing are closely related to images. As Berger states, "Every image embodies a way of seeing. . ... Yet, although every image embodies a way of seeing, our perception or appreciation of an image depends also upon our own way of seeing" (1972:10). What Berger says of image relates to pictorial representations such as paintings and photography. Indeed, every piece of a painting or photograph (and other visual arts as well) embodies individual ways of seeing, which in turn influence persons' perceptions. The same can be true of tourist images. When these are pictorial or other visual representations (MacKay and Fesenmaier 1997), such as images in brochures, they are in effect the embodiment of tourist ways of seeing, quite different from other ways of seeing, such as political business or scientific ways. When images appear as mental representations (e.g., impression, imaging, expectation, etc.), these mental images will shape people's ways of seeing when they travel. Individuals may be open to certain sights that match the images carried in their minds, but indifferent and blind to other sights that do not suit their stereotyped images. For example, windmills, tulips, and prostitutes in windows in the red-light district of Amsterdam may be a tourist image that people have of Holland. As tourists they may be keen to discover these sights, while remaining indifferent or blind to other sights that are more important components of what Holland actually is. Tourist ways of seeing and tourist images are therefore interrelated.

The tourist way of seeing has a number of characteristics. First, it is apoliticized seeing. It shows little concern for local polity. Second, it is a decontextualizing seeing. Tourists usually see only tourist sights and attractions and the social context in which these sights appear is usually ignored. For instance, when foreign tourists enjoyed themselves within the insulated tourist spaces of the Philippines during its rule by Marcos, they were relatively indifferent to the totalitarian régime there. Third, it is a simplifying seeing. Tourists stay in destinations temporarily and only gain a superficial impression of the places visited. The culture and people are reduced to a tourist impression of a space that consists of a few sights, the accommodation, and people such as waiters and tour guides. Fourth, it is an ahistoricizing seeing. A quick and superficial look at destinations does not allow an historical understanding of the local attractions, culture, and people. Fifth, it is a romanticizing seeing. Tourists romanticize sights that confirm the images presented by the mass media or in advertisements they have seen in their homes. The impression they have of destinations is often the projection of previously held stereotyped and idealized images.

Like artistic ways of seeing, tourist ways of seeing also result from training. True, people do not receive formal training in how to sightsee. However, their eyes are trained by culture, and especially by the mass media. As Urry (1990a) indicates, the tourist gaze is socially and culturally cultivated. Tourist ways of seeing are different from the residents' ways of seeing. Locals view their place of residence in a utilitarian and realistic manner. What is seen by sightseers as the extraordinary, the unusual, and paradise is for them the ordinary, the usual, and paramount reality. Tourists are able to transcend the local residents' utilitarian perspective because they are temporary visitors—they come today and go tomorrow (Simmel 1950). Therefore, tourist ways of seeing can be non-utilitarian, romantic, curious, appreciative, and sometimes even neocolonialist. What underlies this is a desire or thirst for something different, unusual, and extraordinary (Urry 1990a), or a place to which to escape temporarily and engage in intensive enjoyment for a short period of time (Cohen and Taylor 1992; Rojek 1993; Shields 1991). Therefore, what concerns tourists most is not how locals suffer as a result of being "overtoured", but rather how they themselves feel during the tour, i.e., whether they are happy, joyful, served well, and having a good time. The tourist's concern for destinations may be highly egoistic in character. As Thurot and Thurot argue, "modern tourism is more of a narcissistic look at one's own culture than at others" (1983:187). In relation to this, tourist ways of seeing contain a component of desire. People may wish certain scenes or themes to be present in a given destination for the purpose of visual or physical consumption.

Tourist ways of seeing constitute a specific kind of seeing. However, they are implicitly shaped by, or related to, other ways of seeing, including political, ideological, scientific, religious, and artistic seeing. Furthermore, class consciousness, group tastes, and national cultural traditions also exert an influence, be this implicit or explicit. Therefore, it is wrong to argue that tourist ways of seeing are homogeneous. On the contrary, they are segmented because they are related to different backgrounds, tastes, and cultural traditions.

Tourist ways of seeing are usually stereotyped seeing. Before tourists arrive at a destination they have already imaged and symbolically structured the place. Their ways of seeing are thus related to "geographical imagination" (Hughes 1992) and cultural classification. They may not have detailed knowledge of a destination, but they may construct an image of it in terms of certain geographical, political, economic, cultural, religious, and ethnic categories, such as "advanced or developing", "democratic or totalitarian", "capitalist or communist", "civilized or primitive", "Christian or Islamic", "Western or Oriental", "White or Black". It is tourists' own cultures that spell out for them these classifications and teach them how to see in a tourist way.

The Creation of Destination Images in Advertising

Tourist ways of seeing have a significant influence upon the tourism industry. Above all, the latter practices a commercial ideology that presents images and discourses that embody such ways of seeing. As Hughes puts it, from its inception "tourism has been framed by particular ways of seeing that are the product of social construction" (1992:32). The industry tends to build an image that embodies, or is geared to, the ways of seeing that exist in its targeted market. Furthermore, once a favorable image enters potential tourists' minds, it will shape their ways of seeing when they are present in the destination.

There are a number of vehicles of communication used by the industry for image making. These include brochures (Adams 1984; Buck 1977; Dann 1988, 1993, 1996b; Dilley 1986; Marshment 1997; Moeran 1983; Reimer 1990; Selwyn 1993), guidebooks (Barthes 1972), advertising on billboards, in newspapers and magazines, and on television (Bojanic 1991; Britton 1979; Hummon 1988; Laskey, Seaton and Nicholls 1994; Lollar and Van Doren 1991; Thurot and Thurot 1983; Urbain 1989), travelogues (Dann 1996c), postcards (Albers and James 1988; Edwards 1996; Mellinger 1994), souvenirs (Gordon 1986; Shenhav-Keller 1993), and videos (Hanefors and Larsson 1993). Of these, the brochure is the most often used. Tourism brochures employ an underlying visual code that structures and organizes image making. This code can be termed "visual inclusion and exclusion" (or "visual selection and avoidance"), which means that certain visual components are included and highlighted, while others are excluded in order to create a favorable image.

Underlying inclusion and exclusion in image formation is a commercial ideology, that is, a system of discourses which distort reality in the name of profit-making. In so doing, the industry engages in a visual ritual whereby features that are culturally desired are highlighted, amplified, and strategically placed. Conversely, what is socially onerous and culturally disliked is avoided and hidden from public view. Tourism advertising, like all propaganda, is a major vehicle of this visual ritual. Destinations may not be able to change their physical and demographic features, but they can alter the ways in which they are visually presented. "For example, a country could try to downplay the importance of attributes on which it is weak and emphasize those attributes on which it has a competitive advantage" (Bojanic 1991:352). Just as "all people ... intensify their own good and downplay their own bad" (Rutledge 1994:213), the tourism industry, via advertising, also visually intensifies the good aspects of destinations and downplays the bad (Urry 1990a:139). Indeed there are what Dann calls "significant omissions" in tourism representations. As he notes,

> the language of tourism, both in its visual displays and in its verbal descriptions, can inform tourists about what to expect at the destination ... not only by what it depicts and says about a vacation, but also by what it leaves out of its pictures and commentaries. In other words, what is omitted may have at least as much influence as what is included (1996a:209).

Even images that are not inherently bad but do not seem desirable in the context of tourism are carefully avoided. As Marshment observes, in Western holiday brochures, "there are no disabled people, no pregnant women, no really fat people, and above all, no black people" (1997:24). It is young, healthy, white, good-looking women and men, beautiful scenery, and comfortable accommodation that are mostly presented in Western beach holiday publicity.

The tourism image is therefore a specific dream image. Several cultural practices contribute to the fabrication of dream images. Fine art, literature, photography, film, television, and tourism are all creators of such images in society. Every society needs not only a realistic image of itself, but also a dream image of a utopia that acts as a reference point. The latter is necessarily a fantasy; yet it offers people spiritual transcendence and a symbolic escape from paramount reality. Above all it is a hope, not a hope for something specific, but a hope in its own right.

What, then, is a dream image? In modern Western culture a typical image consists of Eden, Paradise, Utopia, or Shangri-la. In such a paradise the scenery is beautiful, idyllic, or exotic, and the natives are like Peter Pan—simple, innocent, authentic, and never growing up (Selwyn 1993). From a structural sociological perspective, such an image is the result of the ambivalence of modernity. It is the dark side of modernity that triggers off a contrary and ideal image of holiday utopia to which people can escape (Cohen and Taylor 1992). A tourist image is therefore a socially and culturally constructed utopian image.

However, unlike religion, which offers a prospect of paradise that stands in sharp contrast to earthly existence, tourism actually takes people to paradises in this world. A tourist image is thus constructed out of two resources. One is a general dream image held by people about a worldly paradise that is idyllic and beautiful. The other comprises the physical attributes of places which match this way of seeing, including tropical, palm-fringed islands surrounded by golden sand and clear blue sea. The Caribbean is a typical example of a region that is symbolically transformed into a tourist paradise in such imagery. Relatedly, Wilson (1994) has presented a case study of how the image of the Seychelles has been transformed into one of a paradise in the process of touristification. From the mid 19th century up to the end of the 1950s, white Europeans depicted the Creole population of the Seychelles as "uncultured, idle, drunk, lying, promiscuous, thieving, and superstitious" (Wilson 1994:767). However, with the advent of tourism in the late 1960s and early 1970s a completely different image of the Seychelles emerged in European literature, that of a "Garden of Eden" (Wilson 1994).

As noted above, nothing can be done to change the physical and environmental aspects of a destination, such as the weather. What can be changed is the image of a place in terms of tourist ways of seeing. Thus, even locations that are visually "ordinary" and lack the spectacular landscape or scenery associated with an earthly paradise, can still be symbolically transformed into "extraordinary" destinations (Hummon 1988). This is precisely what advertising often does. "As a ritual text, tourist advertising is involved in a symbolic transformation of reality, remaking ordinary places—from New York to Iowa—into extraordinary tourist worlds" (1988:181). Informing this symbolic transformation of place are tourist ways of seeing. Through visual inclusion and exclusion, by emphasizing certain visual elements and avoiding others, ordinary reality is transformed into an extraordinary dreamworld.

There are a few principles guiding this symbolic transformation of reality: beautification, romanticization, idealization, mystification, and feminization. First, brochures usually select pleasant scenes and omit ugly ones in order to build a beautiful image of the destination (McCrone et al 1995). Second, certain sights that may not be particularly beautiful or spectacular are highlighted because they represent romantic and "idealized images" (Grow 1994). Examples include relics and the primitive. As these sights are romanticized and idealized, they are turned into the beautiful. Third, *mystification* is also a dominant theme in brochures (Britton 1979; Hughes 1992). This is usually applied to poor Third World destinations, for which "Mystification masks reality. The exotic is amplified, usually in the direction of Eden" (Britton 1979:321). In order to maintain such a mystical image, "urban and industrial images are carefully avoided" (1979:321). In addition, wretched poverty, including the image of skinny children dying of hunger, is avoided in order to satisfy tourists' idea of the noble savage (Bruner 1991; Silver 1993). In essence, this mystified image works through a wider framework of "qualitative contrast" (Saram 1983:92) between the image of modernity and the image of non-modernity (the timeless Other, the exotic, and the primitive). The mystified image of Third World destinations acts as a reference point for modernity. Finally, a relatively ignored theme in brochures is the feminization of the image of tourist-receivers (Graburn 1983b; Pettman 1997). In one sense, if overexotic sights cause a feeling of danger, threat, or cultural shock, then the feminization of the host people is a strategy to reassure tourists and reduce the sense of threat. As a result brochures usually avoid the threatening image of the male, and when males do appear they are either children or elderly (signifying the exotic), or else demasculinized as friendly waiters and barmen. Thus, the host population as a whole is feminized (Graburn 1983b) and the "destination becomes a woman" (Dann 1996a:127) in the sense that it is represented only through beautiful scenery and sensuous girls. In another sense, the feminization of the host population increases the erotic ambience and sexual lure of the destination, for "the female body aligns with nature, receptivity, the material or sexual" and thus exists "for men's gaze or use" (Pettman 1997:95). The "stunting of sexuality" in advertising helps promote the sale of products (Moog 1994), and the same applies to the promotion of destinations. Furthermore, brochures not only feminize hosts, but also select, highlight, and amplify the erotic image of female tourists. In brochures, female tourists are often dressed only in swimsuits.

More importantly, they are single, beautiful, tanned, and white (Marshment 1997). There are "no comparable images of men"; it is the image of women in bikinis that functions to signify pleasure in general in beach holiday resorts "because of the strong conventional associations between femininity and aesthetic and sexual pleasure" (Marshment 1997:20).

As a result of these characteristics, the images of destinations in advertising are necessarily distorted and unreal, containing fictional and illusive elements (Britton 1979; Dann 1996a, Hummon 1988; Marshment 1997; Mellinger 1994; Rojek 1997; Silver 1993; Telisman-Kosuta 1989; Wilson 1994). This bias is not simply a moral issue of deliberate cheating and seduction, but rather a sociological question as to why tourists apparently like to be cheated and seduced. When discussing advertising in American life, Boorstin states that, "The deeper problems connected with advertising come less from the unscrupulousness of our 'deceivers' than from our pleasure in being deceived, less from the desire to seduce than from the desire to be seduced" (1964:211). Although he is referring to commercial advertising in general, his observation also applies to tourism advertising. For Boorstin, the advertizer's art is largely one of "making persuasive statements which are neither true nor false" (1964:214). Similarly, images in tourism advertising are not so much true or false, as persuasive. The deeper mechanism of this suasory power consists largely of a wider range of the mass media that creates specific tourist ways of seeing in culture. Images in brochures are seducitive not because advertisers are liars but rather because tourists are culturally trained to be seduced in this way. The production of distorted images requires the complicity of the culture in which tourists find themselves. In other words, tourist images, regardless of whether they are false, are socially constructed (Rojek 1997).

The images of tourism can be related to myth in a Barthesian sense (by contrast Selwyn, 1996a relates image to myth in a Levi-Straussian sense). According to Barthes (1972), a myth is a sign, not a sign in general but a second-order semiological system. For example, the sign of a rose functions as a myth, in that it acts as the signifier of love or passion. Moreover, this artificial relation between the signifying sign of a rose and signified love is naturalized, and hence implicitly justified. As Barthes states, "myth hides nothing: its function is to distort, not to make disappear", and "myth is neither a lie nor a confession: it is an inflexion" (1972:121,129). A myth is naturalized and legitimized inflexion.

Touristic images are linked to myths in two ways. First, in brochures and other mass media, images are constantly turned into myths. For example, the image of a tropical island is converted into the myth of paradise and escape, the image of a camel into desert, the image of Arabs into the mystic Orient, and the image of elephants or native

Thais into the exotic. Second, images are also created by appealing to myths relating to destination areas, myths which are fostered by literature and other mass media. Thus, for instance, to market rural tourism in Britain the industry may resort to the long existing myth of the British countryside and cottages. The myth of the Lake District produced by English literature can also be exploited by the industry. In effect, the image of the Lake District presented in brochures and the mass media is less about the physical attributes of the region, and more about the place myth of English romanticism relating to that district (Urry 1995:ch.13).

The Seduction of Destination Images in Advertising

The twin goals of advertising are to inform and persuade (Kaldor, quoted in Dyer 1982:6–7). This involves the dilemma of honesty and hyperbole. According to Gartner (1993), there are three criteria that can be used to judge image-formation agents: credibility, market penetration, and cost. With regard to credibility, commercial advertisements are seen as less credible than non-commercial sources of information, such as the news. Nevertheless, advertising is still a major agent in image formation because of its high market penetration and wide accessibility. No matter how advertising employs its skills, from its inception it aims to attach a distinctive image to commodities, especially homogeneous commodities such as beer, soap and cigarettes (Boorstin 1964:198). "Image communication" (Gold 1994) is of particular importance in contemporary advertising. "As a system of signification, advertisements compose connections between the meanings of products and images" (Goldman 1992:5). Creating an image and stimulating sales are the two principal functions of advertising (Lollar and van Doren 1991:623). Advertisements are thus an important vehicle for producing idealized images of commodities.

Without exception, tourism advertising also involves the production and communication of destination images. As the sale of places (or products) takes place well ahead of tourists presenting themselves at specific sites, an impressive and convincing image is extremely important in attracting people to buy and consume the products of tourism. The power of destination images to induce is thus highly significant in the economy of tourism in destination areas.

Despite the fact that many potential tourists may well know that the images presented in advertising are distorted, the effects of such advertising are obvious (Bojanic 1991). Why, then, are people seduced by images that they know to be unreal? What makes images seductive? Where exactly resides power of images to induce? The answer seems to lie in the cultural codes that underwrite both the encoding and decoding of images in image communication (Ashworth and Voogd 1990). As to what these cultural codes comprise, certain intersubjectively shared cultural ideas, values, appeals, and ideals (i.e., the stock of values) constitute cultural codes that inform advertisers' image encoding and consumers' image decoding. Both advertisers and consumers have resort to the same sources of ideas, values, appeals, and ideals when encoding or decoding images. Dyer conveys this idea very clearly:

> The advertiser employs language, images, ideas and values drawn from the culture, and assembles a message which is fed back into the culture. Both communicator and receiver are products of the culture—they share its meanings (1982:13).

Thus, advertising does not create new values; it reworks cultural ideologies and values, and produces new meanings out of them (1982:129–130). Advertising tends to appeal to a few basic attitudes and feelings that are associated with certain idealized images of the good life, including: happy families; rich, luxurious life styles; dreams and fantasy; successful romance and love; important people, celebrities, and experts; glamorous places; success in career or job; art, culture, and history; nature and the natural world; beautiful women; self-importance and pride; comedy and humor; childhood (1982:92). Therefore, the images employed in advertising are not themselves inducing; they are seductive because they represent what people already hope and desire. They are encoded in advertising in integrative codes, which include certain styles (humor, narrative, mood, and design) that are impressive, appeals (rational, worry, sensual, testimonial), and values that are cherished by people (Leiss, Kline and Jhally 1997:265–71). The power of tourist images to lure is thus derived from what they convey and represent, that is, socially and culturally sanctioned appeals, goals, and ideals. As Watson and Kopachevsky state:

> Tourism, like other commodities, is packaged for exchange by advertising, much of which appeals to people's deepest wants, desires and fantasies (often sexual), and is anchored in a dynamic of sign/image construction/manipulation (1994:649).

Analytically speaking, there are two fundamental kinds of touristic image: paradisiac images and totemic images. The former represent paradises to which people can escape. The latter include images of totemic national attractions (such as heritage) to which people pay tribute and respect. Paradisiac images are romanticized images that imply that tourism is an escape from the constraining daily world to a romantic dreamworld. Totemic images are sacralized images that imply that tourism is a civil religion, a system of quasi-religious beliefs and rituals; just as people salute the national flag and hold parades, coronation ceremonies and international sporting events, so people salute national heritage and attractions as national totems. The cultural codes that underlie paradisiac images are derived from industrial and metropolitan cultures. As a result, the features that are qualitatively contrasted to industrial and metropolitan cultures, such as the exotic, the pastoral, the primitive, and the curious, are defined as components of tourist paradises. On the other hand, the cultural codes that underwrite totemic images are related to nationalization and national identity. Consequently, heritage, which performs the function of social solidarity and evokes national feelings of pride, is regarded as a sacred and totemic attraction which people honor and travel to as pilgrims. While romanticization plays a major role in the encoding and decoding of paradisiac images, sacralization underpins the encoding and the decoding of totemic images. Although these two codes are analytically separate, in reality they can be mixed with one another.

Therefore, the power of tourist images to induce is related to the wider contexts of culture and society. In more general terms, these images reveal "another world" in response to the ambivalence of modernity. They represent romanticized and sacralized responses to constraining and profane reality. If people cannot overcome the limits of daily reality in material terms, they tend to overcome them in ideal terms, through the romanticization of places to which they can escape, and through the sacralization of national symbols in which they search for meanings.

Touristic images are indicators of the ambivalence of modernity. The seduction of both paradisiac and totemic images must be understood within the context of modernity. On the one hand, tourist images, as idealized images, are the expression of dissatisfaction with the dark side of modernity. On the other hand, they are the embodiment of consumer power. The world of tourism is not only beautiful and attractive but also accessible and available in economic terms. These images are personalized to the extent that a targeted individual can afford to chase them.

Two kinds of "other world" are represented by paradisiac images and totemic images respectively. The former is a tourist paradise; the latter comprises touristic totems. Both of these "other worlds" are constructed in relation to daily and profane reality. A tourist paradise, as represented in paradisiac images, offers a world of pleasure in which people are allowed to release their suppressed desires and undertake roles that are the inversion of their everyday ones (Table 7.1). On the other hand, national totemic attractions, as represented in totemic images, provide access to a sacred world that is superior to the profane world, one in which people can periodically renew their social identity (Table 7.2).

Tourist images are thus symbolic worlds that are superimposed upon the daily world. In this sense, and as MacCannell (1973) argues, tourism is a functional substitute for religion. Tourist images give people the hope

Table 7.1 The Tourist Paradise Represented in Paradisiac Images

Daily Reality	Tourist Paradise
Ordinary world	Extraordinary world
Metropolitan	Pastoral
Industrial	Aesthetic
Work ethic	Fun ethic
Obligations	Escape
Home	Away
Routines	Change
Familiarity	Novelty
Utility	Beauty
Instrumental	Ludic
Reason	Sensation
Rationality	Feelings
Order	Carnivalesque
Self-constraint	Spontaneity
Bread-winning	Consumption
Logos	Eros

of happiness elsewhere in a similar fashion to religion promising people salvation in paradise. As Boorstin puts it: "God makes our dreams come true. Skillful advertising men bring us our illusions, then make them seem true" (1964:212). As such, tourist images represent "another world" created by advertising, just as religion does. Durkheim relatedly claims that "A society can neither create nor recreate itself without creating some kind of ideal by the same stroke" (1995:425). This is one of the functions of religion. Similarly, tourism can assume such a function by creating ideal images as reference points for daily reality. Such ideal images are believed to be real, just as paradise is believed to be real by religious devotees.

Table 7.2. The Touristic Totem Represented in Totemic Images

Daily Reality	Tourist Totems
Profane world	Sacred world
Modernization	Heritage
Dynamic	Conservation
To have	To be
Goods	Meaning
Materialism	Idealism
Daily	Ritual occasions
Organic solidarity	Cultural solidarity

In summary, image building is an important strategy in economic, social, cultural, and political life under modernity. Alternatively stated, modernity involves symbolic representations which constitute a symbolic environment and symbolic seduction. Under the condition of modern capitalism, image packaging is an integral part of commodity production. As tourism also involves commodification, it necessarily includes image building in its production. However, image creation involves wider social and cultural processes, ways of seeing, and values held in society. For images to seduce effectively they must work on certain shared values that people hold about cultural ideals and goals. Therefore, in order to understand the role that images play in tourism, it is necessary to consider them in relation to the wider contexts of society and culture.

Chapter 8

The Lure of Discourse

In the previous chapter, tourist images were examined in relation to the wider contexts of society and culture. However, the formation of images also involves a process of discourse, which exists in advertising, the mass media, education, and word-of-mouth, and which influences tourists' motivations and choice of destinations. In *The Language of Tourism*, a work based on a sociolinguistic perspective, Dann states:

> Via static and moving pictures, written texts and audio-visual offerings, the language of tourism attempts to persuade, lure, woo and seduce millions of human beings, and, in so doing, convert them from potential into actual clients. . . . Thus, since much of this rhetoric is both logically and temporally prior to any travel or sightseeing, one can legitimately argue that tourism is grounded in discourse (1996a:2).

Crawshaw and Urry note that

> Different gazes are "authorised" by different discourses. Examples include the discourse of *education* which conditioned the experience of the European Grand Tour, that of *health* which defines a type of tourism whose aim is to restore the individual to a state of physical well-being, and the discourse of *play* which surrounds what can be called "liminal" tourism (1997:176).

Of course tourists can often feed back into this discourse or even challenge it by inventing their own discourse (e.g., word-of-mouth). Yet the influence of discourses upon potential tourists is undeniable. People choose to travel somewhere partly because they are, consciously or unconsciously, persuaded or lured by a certain discourse to which they have been exposed for some time. What travelers see is what is what they expect to see, for the discourse has already informed them where to go and what to see (Dann 1996a; Frow 1991). Therefore, the study of discourse helps deepen the understanding of tourism, particularly as it relates to tourist choice and taste.

This chapter deals with discourse from a sociological perspective. Rather than employing discourse analysis and content analysis, this chapter treats discourse as a social fact in a Durkeimian sense, and examines this fact in terms of the relationships between discourses and other social facts. It is argued that the discourse of tourism embodies a collective taste within a given culture or social class. Thus, it is viewed as a discursive representation of collective preference. In this sense the discourse of tourism is an objective social fact, worthy of *macro*-sociological study. This chapter is composed of four sections. The first discusses conceptual issues. The second deals with the issue of how discourse functions as the legitimization of travel and tourism. In the third section the question of how discourses legitimize taste is considered. Finally, the fourth section elucidates how discourses act as agents of seduction.

Conceptual Issues of the Discourse of Tourism

First of all it is necessary to define "discourse". According to Foucault, "discourses are composed of signs; but what they do is more than use these signs to designate things. It is this *more* that renders them irreducible to the language (*langue*) and to speech" (1989:49). Clearly discourses are not the same as *langue* and speech, since there are both spoken and written discourses. "Discourse" is sometimes distinguishable from "text" (Stubbs 1983:9). Moreover, a discourse not only consists of words and statements, but can also contain elements of music, images, or pictures (Cook 1992:1). For instance, as Postman puts it, "on television, discourse is conducted largely through visual imagery, which is to say that television gives us a conversation in image, not words" (1985:7).

A discourse is a system of statements. The basic unit, however, is not a single statement but an interrelated system of statements, which speaks out through its totality rather than through a constituent part. In this sense, discourse is a practical process and a structure of thoughts and feelings. Unlike ideology, discourse has little to do with truth or error, rather it is a discursive formation that embodies collective consciousness. In a Foucaultian sense, discourse is a discursive practice that delineates, defines, or justifies something, while simultaneously avoiding or excluding others. It is a practice whereby people construct reality, that is, the object of discourse.

There are various vehicles of discourse and discourses can be classified in terms of these vehicles. There are oral, written (letters and diaries), typographical (books, newspapers, magazines), audio/visual (radio, music, films), and electronically-mediated discourses (television, video, the internet, CD-ROM) (Dann 1996a). Alternatively, one can speak of media, classroom, advertising, political, civil, legal, business, religious, scientific, or tourism discourse. Discourses are thus characterized not only by their vehicles but also by their content.

Tourism discourses are a particular kind of discourse whose carriers include word-of-mouth communication, letters, travelogues, travel books, brochures, souvenirs, advertisements in the media (billboards, newspaper, magazines), holiday programs on television, travel information on the Internet, video tapes, and CD-ROM. In addition to these vehicles, there are wider processes of communication which also involve or imply tourism discourses, such as school textbooks, classroom teaching, films, paintings, and so on. Tourism discourses have many kinds of contents. Examples include the justification of travel, accounts of tourist experiences, portraits of attractions, descriptions of the local customs, culture, and people in destinations, reports about accommodation, transport, and other necessary services, and judgements about tourism fashions. The list of possible contents goes on and on. However, only a few selected themes are highlighted here: the legitimization of travel, the legislation of tourist taste, and the power of discursive persuasion and seduction. Tourism discourses are opinion makers and leaders. They influence, consciously or unconsciously, potential tourists' motivations and choice of destination. As a discursive environment to which people are frequently exposed, discourses function as seducers or hidden persuaders. Furthermore, they train potential tourists in matters of taste: they instruct them where to go, what to see, indeed the art of how to travel; they create the expectations and anticipations of experiences for potential tourists.

Tourism discourse is not the same as a "marker", which is always linked or attached to a particular object or sight, such as markers of moon dust (MacCannell 1976), the battlefield of Waterloo, and the birthplace of Shakespeare. However, markers are an integral element of discourse. While markers are crucial to attractions, in themselves they cannot, however, determine whether an attraction is popular or not. What determines this is, to a large extent, discourse. Certain attractions that used to be popular may no longer be so, such as seaside holiday resorts in Britain, including Scarborough, Blackpool, and Brighton (Urry 1990a:16–39). At the same time, certain sites that were seen as ordinary in the past may be popular today. For instance, Bradford, a British industrial city which was formerly avoided by one and all, has been turned into an increasingly popular destination. Such changes in the popularity of attractions are not completely due to a change in their objective features, but rather or at least partly, by a change of discourse which reflects a modification of tastes and images.

Tourism discourse involves a particular orientation towards the world. Generally speaking there are three ideal-typical orientations: the religious, the scientific, and the touristic. Religious orientation emphasizes "paradise" at the expense of "this world". Scientific orientation stresses "this world" at the expense of "paradise". Touristic orientation high-lights the mixing of both "paradise" and "this world", i.e., paradise in this world. Tourism is thus a mediation between religion (illusions and beliefs, "hot" orientation) and science (observation and discovery, "cool" orientation). Tourism discourse, which informs orientation, also mediates between religious discourse and scientific discourse. Informed by tourism discourse, tourism lies between the profane and the sacred. Tourism is both the *idealization* of "another world" (like religion) and an *exploration* of the world as it could be (like science).

Different discourses underwrite different forms of tourism. It must be noted that a particular kind of tourism may involve several discourses at the same time rather than just one. However, not all have the same influence. For example, a discourse of "Exoticism" may exert more influ-ence on Oriental tourism than do other discourses. Thus, it is necessary to identify the specific discourse that dominates a specific type of tourism during a particular period of time. Moreover, some non-tourism dis-courses may compete with tourism discourse and may exert a certain influence upon tourist behavior. For example, the scientific "ozone–mel-anoma" discourse, which focuses on the risk of skin cancer from over-exposure to the sun, has increasingly competed with a "sun-is-fun" discourse, which focuses on pleasure and hedonism in the sun, leading to a reduction in the number of beach holidays. Conversely, the "body-culture" discourse, which focuses on the symbolism of tanned skin, still entices people to the beach for their holidays, although with judicious use of sun-screen lotions (Coupland and Coupland 1997). Thus, tourism dis-course must be considered in relation to other discourses.

Outline of a Macro-historical Analysis of Tourism Discourse

The tourism discourse arose simultaneously with tourism itself. It can be argued that the forms of tourism are to a large extent determined by tourism discourses and that a specific form of tourism is accompanied, or anticipated, by a particular discourse. In effect, it is the appearance of discourse, rather than the emergence of the tourism experience, that more properly signifies the beginning of modern tourism. In short, mod-ern tourism began with a collective consciousness of the tourist's needs, which was represented in various discourses.

A Brief Review of Western Tourism Discourses

The concrete contents of discourses, however, vary in different histor-ical periods. Roughly and ideal-typically speaking, there are several identifiable types. In the period of the Grand Tour, an age that signaled the start of modern tourism, a discourse of *education* dominated the practice of tourism (Pimlott 1976), and the original sites of the Renaissance in continental Europe, such as Italy, became attractions for young English aristocrats (Towner 1985; Turner and Ash 1975). Later on a discourse of *health* appeared. The popularity of taking the waters in the spa and seaside resorts of Britain is one example (Hern 1967; Pimlott 1976; Swinglehurst 1974:27–28; Urry 1990a:16–39). While the discourses of education and health still existed in the latter half of the 18th century, a discourse of *romanticism* also emerged. This informed the practice of nature and landscape tourism during that period and literary tourism later on (Andrews 1989; Jasen 1991; Ousby 1990; Squire 1988). Thomas Cook's tours seem to mirror a discourse of *rational leisure and recreation*, which arose in relation to temperance and attempts to curb drunkenness against a backdrop of boredom in the age of industrialization (Swinglehurst 1974:20). Thomas Cook invented a rational way of organizing tourism, which has had a great deal of influence upon the industry ever since. With the spread of colonialism in the 19th century a discourse of *exoticism* appeared (Todorov 1993), which was reflected in the travel literature. This lasted until the 20th century post-colonialist period (Silver 1993). In association with this discourse, international tourism, particularly ethnic tourism, was fostered. In the post-war period the discourse of *escape or paradise* loomed large (Cohen and Taylor 1992). Linked to this was the emergence of mass tourism and the growing popularity of coastal resorts that offered natural attractions and sexual opportunities, denoted by the "4Ss" (sun, sand, surf, and sex). Nowadays, a discourse of *pluralism*, which is related to segmentation of the market, reigns in postmodern cultural life. Connected to this has been the diversification of attractions. Since tourists are no longer seen as homogeneous, attractions have undergone differentiation, suiting the different needs of tourists who pursue their lifestyles and tourist identities in various ways.

An Analysis of Western Tourism Discourses

Human beings travel. This is a universal phenomenon. For gypsies, travel is the basic way of life. For settled peoples, travel is not a necessity, although it is sometimes regarded as indispensable. Unlike nomads and gypsies, travel for settled peoples is an unusual event, a break from the daily routine. Such an unusual event often entails justification and legitimization, for travel implies extra spending and risk. Individuals must have a reason or purpose to travel.

In premodernity the justification of travel usually lay in its utilitarian functions and purposes. The main model for non-utilitarian travel in the Christian Middle Ages was that of pilgrimage (Stagl 1995; Vukonić 1996), which, if seen from a functionalist perspective, was also utilitarian in the sense that religion was fundamentally functional to people's lives. However, according to Stagl (1995), towards the end of the Middle Ages pilgrimage quickly lost its legitimacy, for the secular components of travel increasingly militated against the religious domain. "About the year 1550 pilgrimage had ceased to be a plausible justification for travel. A new legitimation was needed", and the new legitimation "was found in *education*" (1995:47). Thus, the surface of the earth was increasingly explored from the 16th to the 18th centuries and travel was regarded as an important way of learning and obtaining knowledge about the world. The flourishing of travel writing during that period is evidence of the fact that travel writing was seen as a legitimate reflection of a modern conception of the world (Stagl 1995). It can be argued that empirical observation and discovery through travel were elevated to a modern empirical epistemology by the English philosopher Francis Bacon. Such epistemological or scientific travel was epitomized by Charles Darwin's voyage to South America and the Galápagos Islands, which resulted in his evolutionary theory of the origin of species.

Thus, generally speaking, tourism discourses in early modernity tended to emphasize the utilitarian functions of tourism. By contrast, under late modernity they tend to stress the non-utilitarian functions of tourism, such as pleasure, entertainment, recreation, and enjoyment. They focus on feelings, hedonism, escape, consumption, and lifestyle. Such a shift of focus is by no means an accident. It implies a fundamental change of legitimacy in the motives of travel.

The earlier forms of tourism were, to a certain extent, bound to a work ethic, a culture of production, and rationalization. Thus, discourses tried to legitimize travel and tourism in utilitarian terms and were consonant with reason and rationality in an age of reason and industrial capitalism. That tourism should be productive or beneficial to either the body (health) or the mind (knowledge) became rational criteria in the judgement of touristic legitimacy. Non-productive leisure and travel could claim no such legitimacy. Even if non-productive components of travel such as pleasure and sensual excitement were part of the motive for travel, they had to be masked in productive terms. Thus, in Thomas Cook's organized tours, non-productive components such as pleasure certainly existed and became attractive features. However, they were discursively concealed under productive and utilitarian terms such as temperance or the rational control of leisure (Swinglehurst 1974).

As a reaction to this rationalist discourse, there gradually appeared a romantic discourse from the latter half of the 18th to the 19th century.

Correspondingly, a different kind of legitimacy came into being, which focused on feelings, passions, and pleasure, and transcended utilitarian and pragmatic considerations. As a result, picturesque landscapes, the countryside, the past and its relics, as well as the exotic, were appreciated for their own sake. By contrast, industrial and urban environments were denigrated. In relation to this romantic discourse, "escape" became a dominant theme in travel and tourism, and has remained so ever since. Romantic discourse is the discourse of counterindustrialization, of nostalgia for the past.

Finally, with the emergence of the postmodern condition, and especially the advent of hyper-consumerism, tourism discourses seem to become diversified, reflecting a pluralism in touristic legitimacy among Western people. Here lifestyle discourse appears to be a central kind of tourism discourse. It is claimed that whatever tourists seek, they are in fact pursuing their own distinctive lifestyles and laying claim to their own particular identities. Lifestyle discourse is related to the emerging touristic consumer culture under the postmodern condition.

In sum, tourism discourses, as the legislation of touristic legitimacy, embody collective consciousness, tastes and preferences, and can thus be collectively seen as a social fact. This social fact is closely related to other social facts, such as wider cultural processes, the collective consciousness, and collective values.

Tourism Discourse as the Legislation of Taste

Ousby (1990:5) claims that patterns of travel are related to movements in taste. Taste is usually legislated and expressed in discourses. If there is rivalry between different tastes, then they are supported by competing discourses. The discourse of the traveler versus that of the tourist is one of the most typical tourism discourses on taste in the West.

Originally there was no significant difference between the word "traveler" and the word "tourist". However, during the first half of the 19th century a distinction came to be made between the two, with "traveler" being used in a commendatory sense and "tourist" being a derogatory term (Buzard 1993). Both traveler and tourist were thus embodiments of different cultures of taste—high culture in the case of the former and low culture in the case of the latter (Rojek 1993). In 1869 Henry James, a man of high taste, wrote that "tourists are 'vulgar, vulgar, vulgar'" (quoted in Pearce and Moscardo 1986:121). Those of high taste objected to being labeled "tourists" partly because the means of transport used by tourists, and inclusive package tours such as those offered by Thomas Cook, deprived travelers of initiative and the spirit of adventure. Thus, John Ruskin complained: "Going by railroad, I do not consider as travelling at all; it is merely being 'sent' to a place, and very little different from becoming a parcel" (quoted in Boorstin 1964:87). In 1865 an article in *Blackwood's Magazine,* written by a British consul in Italy, attacked Thomas Cook's way of tourism. He wrote that the cities of Italy were now

> deluged with droves of these creatures, for they never separate, and you see them forty in number pouring along a street with their director—now in front, now at the rear, circling round them like a sheepdog—and really the process is as like herding as may be. I have already met three flocks, and anything so uncouth I never saw before, the men, mostly elderly, dreary, sad-looking; the women, somewhat younger, travel-tossed, but intensely lively, wide-awake, and facetious (quoted in Boorstin 1964:87–88).

This kind of discourse of condemnation of the tourist in favor of the traveler was shared by many intellectuals in the 19th century and continued throughout the 20th century, as exemplified in the book *Traveler's Quest: Original Contributions towards a Philosophy of Travel*, edited by M. A. Michael (1950a). In the introduction, Michael argues that holidays bought from travel agencies constitute "bogus travel" and have nothing to do with "real travel". He defines the "ideal" traveler as

> one who in the first place has set out in search of something, definite or indefinite. He may have a concrete aim, or just a vague longing, but his journey is a quest. In his travels he must enjoy absolute liberty and independence (1950b:5).

More concretely, for Michael ideal travelers are characterized by six factors. First, they have only a vague goal and not well-planned destinations. Second, there is no time limit on their travel—they have plenty of time. Third, and relating to the second factor, they must have sufficient income to guarantee their independence. Fourth, they are open minded, "able to talk with people encountered whilst traveling" (1950b:7), and even able to speak the language of the host community. Fifth, they have a spirit of adventure and are always trying to discover something new or unknown, urged on by childlike curiosity. Finally, they possess the knack of how to see things. However, such an ideal traveler is rarely possible for obvious reasons, such as constraints of work and other obligations, or insufficient time and income. Nevertheless, the ideal traveler functions as an ideal representation of the highest taste of travel and against which various forms of bogus travel can be criticized.

For Michael, travel should not take the form of a package tour, nor should it use advanced means of transport that may constrain a traveler's liberty and independence. Modern means of transport such as "the railway, the motor car and the airplane are convenient means of approach, nothing more. The travel begins where transport stops" (1950b:4). He continues,

> if you are truly to travel in the meaning of that word … you must go on foot
> or employ the services of dog or horse, or boat…. All other travel [by
> mechanical means of locomotion, airplane, etc.] is more or less bogus, in
> essence just touring (1950b:18).

Moreover, in contrast to a package tour, from which people only gain
insulated experiences from contrived attractions, travel offers real contact with nature:

> How can you travel if you cannot hear the songs of the birds, the ripple and
> chatter of water, the music and wild poetry of the wind, the barking of geese
> or the hot cooing of wood-pigeon? How can you travel if you cannot stop
> when a hare or fox lollops out of the hedge, or when you see a bird or animal
> freeze a few yards away? How can you travel if you cannot smell the scent of
> the hedge-rows and trees, the fragrance that comes on the air, or if you
> cannot feel the weather? How can you travel if there is no silence and no
> solitude? You cannot (1950b:19).

This is a typical example of a discourse that legislates the taste of travel, a
form of travel that should be preferred to that of tourism. Travel is
defined as an experiential art, a high taste of the romantic appreciation
of nature, as opposed to vulgar tourism, whicht is devoid of genuine
experiences.

Another typical example is Boorstin's comment on the dichotomy
between the traveler and the tourist. In a chapter entitled "From traveler
to tourist: the lost art of travel", Boorstin conducts a nostalgic critique of
mass tourism in terms of the genuine travel of the past and laments "the
decline of the traveler and the rise of the tourist" (1964:84–85):

> The traveler, then, was working at something; the tourist was a pleasure-
> seeker. The traveler was active; he went strenuously in search of people, of
> adventure, of experience. The tourist is passive; he expects interesting things
> to happen to him. He goes "sight-seeing"…. He expects everything to be
> done to him and for him (1964:85)

Tourism, as a commodity, thus indicates the loss of the art of travel; it
ceases to be genuine travel. Travel involves hardship, adventure, real
experiences, initiative, and discovery. By contrast, tourism is characterized by comfort, ease, lack of risk or false adventure, contrived tourist
sights, and insulated experiences. Thus, Boorstin condemns tourism as a
kind of pseudo-event. His discourse implies appreciation of the high taste
of travel and deprecation of the low taste of tourism in terms of means of
transport, the nature of experiences, the nature of attractions, and the
form of travel, as summarily illustrated in Table 8.1.

Since the 1980s the discourse of the dichotomy between travel and
tourism has been echoed by the discourse of "alternative tourism".
Although the meaning of alternative tourism is ambiguous and it
means different things to different writers, it is clear that alternative

Table 8.1. The High Taste of Travel and Low Taste of Tourism

	High Taste of Travel	**Low Taste of Tourism**
Transport Experience	• Simpler means • Hardship • Risk and adventure • Active quest and discovery • Self-reliance • Contact with local people and culture • Genuine experiences • Exploration-related	• Advanced transport • Comfort and ease • Lack of risk • Guided sightseeing and gullibility • Passivity • Insulation from local people and culture • Contrived experiences • Consumption-related
Attractions Form	• Original and authentic • Independent travel	• Contrived and inauthentic • Package and organized tour

tourism is quite different from mass and package tourism, and that it is regarded as relating to high taste.

It must be noted that tourists may choose certain patterns of travel according to their tastes and personal preferences rather than as a result of discourses. However, their tastes are partly the result of conscious or unconscious exposure to discourses that exist in travel writings, brochures, or the mass media. They dialogue with these discourses, accept a particular one, and finally are socialized into being a particular type of person with a special pattern of taste, which is reflected in their choice of holidays.

Both taste and tourism discourses are related to social class. Therefore, in referring to taste and discourse it must be asked "who is the legislator of taste?" and "who constructed the discourse?" However, there is no single legislator. In reality discourses are plural, each relating to the ideology of each social class (Thurot and Thurot 1983). Usually it is the highest social class that dominates, directly or indirectly, the field of discourse and thus governs the legislation of taste in tourism. Tourism discourse involves elitism. With the support of capital the dominant class upholds "destination elitism", an elitism characterized by travelling to certain destinations beyond the reach of other social classes (Moeran 1983). However, as intellectuals are the direct producers of knowledge and discourses they are a major class of legislators of taste through discourse. For example, a new petite bourgeoisie (e.g., teachers), lacking sufficient capital, may choose certain holidays which are cheap but nevertheless demonstrate their distinctive yet superior cultural taste, such as a sporting holiday or ethnic tourism (Bourdieu 1984; Munt 1994). The discourse of travel versus tourism, as discussed above, embodies the ideology and cultural capital of intellectuals.

In relation to social class, taste can be expressed and legislated through either verbal or tacit discourse, the latter being exhibited in tourists' holiday choices. To put it another way, tourist choices embody tacit discourses in and through touristic objects, thus, the members of each social class assert their own distinctive taste and status through tacit discourse in the choice of objects (accommodation, modes of transport, attractions, and destinations). In *Distinction*, Bourdieu indicates that the choice of holidays is one of cultural preference that expresses certain distinctive tastes of social class. He writes:

> Taste, the propensity and capacity to appropriate (materially or symbolically) a given class of classified, classifying objects or practices, is the generative formula of life-style, a unitary set of distinctive preferences which express the same expressive intention in the specific logic of each of the symbolic sub-spaces, furniture, clothing, language or body hexis (1984:173).

The choice of holidays, along with other cultural practices and choices, is then an expression of a distinctive taste and lifestyle, consonant with class position. The difference in tastes and lifestyles between, for instance, the new petite bourgeoisie, such as teachers, and the old bourgeoisie can be seen in their different cultural practices, which include their choice of holidays:

> the teachers' walking, camping, mountain or country holidays are opposed both to the set of luxury activities and goods which characterize the old bourgeoisie—Mercedes or Volvo, yachts, hotel holidays in spa towns—and to the constellation of the most expensive and prestigious cultural and material possessions and practices—art books, movie cameras, tape recorders, motorboats, skiing, golf, riding or water-skiing—which distinguish the liberal professions (1984:283).

While the members of the petite bourgeoisie, who differ from the old bourgeoisie in matters of taste, tend to choose a relatively inexpensive lifestyle to express their distinctive taste and cultural superiority, the new bourgeoisie, for example private-sector executives, differ from the old bourgeoisie by mixing the "intellectual" with luxurious lifestyles. Bourdieu describes the new bourgeoisie in the following way:

> Having achieved positions of power at an earlier age, more often being graduates, more often belonging to bigger, more modern firms, the private-sector executives are distinguished from the industrial and commercial employers, a traditional bourgeoisie with its spa holidays, its receptions and its "society" obligations, by a more "modernist", "younger" life-style, certainly one that is more consistent with the new dominant definition of the dynamic manager (although the same opposition is found among the owner-employers) (1984:305).

According to Bourdieu, these executives are more "oriented toward the outside world", more "open to modern ideas", and correspondingly "have the highest rate of foreign travel" (1984:305). Their travels include holiday-making, business travel and attendance at professional conferences or seminars. Their lifestyles are "cosmopolitan" in character (1984:309).

Therefore, holiday-making has increasingly gained an expressive function. It is an art, a lifestyle, a choice of cultural activities based on taste and preference, rather than on universal and standardized demand. Holiday-making is one of the areas in which competition for symbolic capital and status takes place. "We are starting to be judged on our leisure persona, rather than our work role" (Turner and Ash 1975:14). Thus, tourism, as well as leisure time, is a free choice and an expression of tastes and preferences.

The legislation of taste involves the appraisal of touristic objects, such as mode of transport, accommodation, attractions, and souvenirs. Tourists with a certain taste will prefer a certain mode of transport, accommodation, and so on because for them the choice of touristic object is the embodiment of tourist taste. For instance, authenticity is one of the criteria used to decide whether attractions are worthy of a visit, or souvenirs are worthy of purchase (MacCannell 1973, 1976; also see *chapter 10*). As far as attractions are concerned, objects become attractions not simply because of their physical features, but at least partly because of people's tastes and preferences, which are reflected in tourism discourses. For example, nature, which used to be an object of fear, becomes an attraction not just because of its physical features, but also because people's orientation towards nature changes, that is, nature fits their romantic taste (see *chapter 4*). With the rise of nostalgic taste, heritage, as the tangible evidence of the past, has also become a popular attraction (Lowenthal 1985). As discourses greatly influence people's taste, attractions are therefore to a certain extent formed on the basis of discourses. As Dann claims, "It is *discourse* which delineates the sight, *language* which informs the tourist what must be seen *before* the journey is undertaken" (1996a:21). Thus, the "pull" of attractions is to a large extent dependent upon the power of discourse to seduce.

Tourism Discourse as the Power of Seduction

Discourse may attach a certain magic to objects. As Poster states:

> The commodity-object in the TV ad is not the same as the one taken home from the store and consumed. The latter is useful but prosaic, efficient but forgettable, operational but ordinary. The object in the ad and in the store display is magical, fulfilling, desirable, exciting. The difference between the two is produced by the TV ad in its communication which constitutes the subject within the code (1990:63).

Thus, advertising discourses on television add an emergent quality to objects, which makes them attractive in a particular way. Discourses are thus an important element in the social and cultural construction of reality. The representation of objects in discourse turns the world of goods into a world of culture. As Postman claims:

> We do not see nature or intelligence or human motivation or ideology as "it" is but only as our languages are. And our languages are our media. Our media are our metaphors. Our metaphors create the content of our culture (1985:15).

In terms of constructivism (for details see *chapter 3*), there is no reality that is independent of an individual's perspectives and interpretations. Interpretations are always carried out in and through language. Therefore, reality is not what "it" is "but only as our languages are". In Gadamer's (1976) terms, language is the ontological condition of interpretation. This is the reason why discourse has power. However, the power of discourse is "hidden" (Fairclough 1989). In discussing the media as a form of power, Fairclough writes:

> It is a form of the power to constrain *content*: to favor certain interpretations and "wordings" of events, while excluding others (such as the alternative wording...). It is a form of hidden power, for the favored interpretations and wordings are those of the power holders in our society, though they appear to be just those of the newspaper (1989:52).

What is said of the media also applies to discourse as the media are discourses. However, constraining content is only one element of the power of discourse. According to Fairclough there are two major aspects of a power–language relationship, namely power *in* discourse and power *behind* discourse. As illustrated by the example of face-to-face discourse, "power in discourse is to do with powerful participants *controlling and constraining the contributions of non-powerful participants*" (1989:46). Three types of constraint can be identified: constraint on content (on what is said or done), relations (the social relations people enter into in discourse) and subjects (the "subject positions" people occupy) (1989:46). Power behind discourse means "that the whole social order of discourse is put together and held together as a hidden effect of power" (1989:55). Thus, the power in media discourse is the embodiment of the power behind media discourse. Media producers exercise power over media consumers "in that they have sole producing rights and can therefore determine what is included and excluded, how events are represented, and ... even the subject positions of their audiences" (1989:50).

In relation to tourism, three kinds of discourse can be distinguished: word-of-mouth (face-to-face discourse), media, and advertising discourses. Word-of-mouth discourse exists among relatives, friends, and colleagues. The subjects of the discourse are familiar, or come to be familiar, with one another. When tourism is referred to, power in the discourse arises among those participants who have visited certain attractions which other members either cannot afford or have failed to visit. The experience of attendance makes the witnesses of these attractions and events the judges of whether they are worth visiting. Hence, they come to be speakers and others come to be listeners in relation to these attractions. Their words about their experiences will have a significant influence upon their listeners. Indeed, in many cases people choose a particular holiday because of friends' recommendations. Within the circle of a close community, word-of-mouth is most convincing and credible. Its inducing power is implicit but significant. However, the extent to which word-of-mouth works is limited (Gartner 1993).

As regards media discourse, what is referred to here is a non-commercial discourse on tourism or touristic attractions. Nowadays, holidays have become an interesting topic and occupy significant space in newspapers, magazines, and television. For example, among the popular programs on British TV are "Holiday" on BBC 1, "Wish You Were Here" on ITV and "Travelogue" on Channel 4. Certain other programs, which are not directly about travel and tourism, include a great deal of holiday content, such as the extremely popular "Blind Date" on ITV. Although "Blind Date" is sponsored by a commercial company (Free Choice), its commercial bias is concealed. The seductive power of these media discourses is considerable, and as they are seen as largely free from commercial bias they are regarded as convincing. Consequently, they tend to become an invitation for participation. The title "Wish You Were Here", for example, suggests conversion from potential tourist into actual holiday-maker. As the target coverage is quite extensive, the degree to which media discourse influences people is also large (Gartner 1993). Media discourse thus has a demonstration effect on the one hand, and brings abut cultural conformity on the other.

Third, advertising or commercial discourse is part of media discourse because advertisements have an intrinsic connection with the mass media. It is due to this association that advertising discourse becomes ubiquitous and increases its market penetration. However, advertising discourse can be separated from media discourse in general for analytical purposes. The former is a commercial agent and contains commercial bias, and is thus seen as less convincing than community or media discourses. As Dyer observes, "Many people would deny that they are influenced by advertisements and regard them at worst as lies, at best as idiot triviality" (1982:72). Indeed the power of advertising to convince is relatively low because an advertisement is an overt inducing agent (Gartner 1993). However, as Dyer notes, although most people might not believe the claims made for a product by an advertiser ... they might find it difficult to resist the more general social image or message presented along with the overt sale pitch—for example, that we can make friends by drinking the right kind of beer, get a boyfriend by using the right kind of shampoo, become a superman to an adoring family by buying the right tin of baked beans, or avoid being a social outcast and guilt feeling if we buy life assurance (1982:72).

Thus, what is central to advertising discourse is not to inform but to persuade, to "create desires that previously did not exist" (1982:6). In so doing, advertising discourse tends to reveal audience deficiency in terms of general ideal images of life or social identity that the advertised items represent. In Williamson's words, "advertisements *intend* to make us feel we are lacking" (1982:8). To achieve this, advertisements not only try to sell things to us but, more importantly, also tend to make these things "mean something to us". "[A]dvertisements have to translate statements from the word of things ... into a form that means something in terms of people" (1982:12). In effect, consumers are persuaded and manipulated into buying a way of life as well as goods (1982:5).

Leiss et al (1997) point out that there are four basic advertising formats: a product information format, a product image format, a personalized format, and a lifestyle format. In the third and fourth formats, advertising discourse tends to create meanings for goods among people in terms of their nature as social beings or their ideal images of life. As a result, discourses have the effect of making consumers feel that they cannot do without these goods.

The advertising of tourist destinations has a similar effect. In advertisements, tourism is highlighted as an idealized lifestyle, a display of social identity, and a marker of status, and touristic attractions are portrayed as not only worthy of visit but also a "must". What is hidden in discourses is the message "wish you were here", which in turn implies that absence from a destination carries a stigma. A tourism discourse is thus, in a sense, a normative discourse. It encourages social conformity and gradually leads clients into the consumer class of "I can't miss it", a new generation of class, as described by Baudrillard:

"Try Jesus!" says an American slogan. Everything must be tried: since man as consumer is haunted by the fear of "missing" something, and kind of pleasure. One never knows if such and such a contact, or experience (Christmas in the Canaries, eel in whisky, the Prado, LSD, love Japanese style) will not elicit a "sensation." It is no longer desire, not even "taste" nor a specific preference which are at issue, but a generalized curiosity driven by a diffuse obsession, a *fun morality*, whose imperative is enjoyment and the complete exploitation of all the possibilities of being thrilled, experiencing pleasure, and being gratified (1988:48–9).

Such consumer consciousness is well exploited by advertisers in their discourses. People may realize that tourism is non-essential consumption when compared with bare subsistence. However, they may not resist the invitation, seduction, or imperative of the normative discourses of advertisements.

In summary, tourism discourses, as part of the social fact of tourism, are related to other social facts and can be considered in relation to the wider contexts of society and culture. Such discourses are the discursive representation of the collective consciousness with regard to "another world" that functions as a holiday destination. Discourses are shaped by concrete socio-economic (such as industrialization) and socio-cultural conditions (such as rationalization, romanticization). They play a role in legitimizing not only travel and tourism, be this utilitarian or hedonistic, but also taste in relation to tourist choice. Discourses employ various vehicles of communication, such as word-of-mouth, the mass media, and advertisements, all of which have the power to seduce. Whereas different forms of media have different degrees of credibility and market penetration, advertising discourses play a significant role in seducing tourists.

Chapter 9

The Lure of Consumption

Under modernity the production of tourism takes the form of commoditization. Universalizing commoditization extends from the production of goods to the production of experiences, including leisure, entertainment, recreation, and tourism. Thus, tourism, as a result of commoditization, becomes a commodity for mass consumption (Watson and Kopachevsky 1994). Relatedly, a touristic consumer culture has emerged (for a review of theories of consumer culture see Corrigan 1997; Featherstone 1991b; Lee 1993; Lury 1996; Slater 1997).

Touristic consumer culture can be considered in two senses. First, in modern capitalist societies there is the cultural orientation towards the marketing and consumption of the goods, services, and experiences of tourism. In this sense, tourism is commoditized and consumed as the end product of hedonistic experiences and enjoyment. Second, the consumption of tourism appears as *symbolic* consumption, which is related to the culture of status differentiation and market segmentation, in which the choice of products not only reflects the social position of individuals (age, gender, occupation, ethnicity, etc.) but also their tastes, social values, and lifestyles. In this chapter it is the first sense of touristic consumer culture that is the principal focus. The second sense, namely, the symbolic consumption of tourism, will be dealt with in next chapter.

In the literature on the sociology of consumption, consumption is sometimes viewed as "mentalistic hedonism" or "imaginative pleasure-seeking", a constant quest for novel products:

> The essential activity of consumption is thus not the actual selection, purchase or use of products, but rather the imaginative pleasure-seeking to which the product image lends itself, real consumption being largely a resultant of this mentalistic hedonism. Using this framework it becomes possible to understand how it is that modern consumption centered upon the consumption of novelty. For modern consumers will desire a novel rather than a familiar product because this enables them to believe that its acquisition and

use will supply experiences that they have not encountered to date in reality (Campbell 1995:118).

According to Campbell, such modern consumerism is best illustrated by reference to tourism. Tourism is essentially both novelty-seeking and imaginative pleasure-seeking. Thus, "if tourism is best understood as a form of imaginative hedonism, then perhaps this might also be the best way to understand modern consumption in general" (1995:119). Therefore, treating tourism as hedonistic consumption helps illustrate the *experiential aspect* of consumer culture in general.

In modern societies tourism, as the commoditization of experiences, leads to a consumerist transformation of the nature of travel. Travel, which was an adventurous experience in the past, has been converted into a form of leisure and a hedonistic experience, and hence integrated into modern consumer culture. The "crude materials" of travel have been processed and refined as the "end product" of tourism. With the aid of advertisements and marketing, this product triggers off a desire for imaginative pleasure-seeking. Symbolic components such as images, representations, and discourses, as discussed in the previous two chapters, are integral elements of the touristic consumer culture, and therefore intangible products can be displayed in, say, travel agencies.

This chapter deals with the experiential (hedonistic) nature of touristic consumer goods and services and the issue of how experiential consumer products are produced and supplied in modernity. There are three sections. The first tackles the issue of how the commoditization of tourism transforms travel into tourism. In the second section, tourism is treated as experience-centered consumer culture, in contrast to goods-centered consumer culture. Finally, the third section analyzes the question of how tourist space, as a consumption site, seduces consumers.

The Commoditization of Tourism in Modernity

The differences between travel and tourism are evident. In the Middle Ages travel was risky, expensive, connected to travail and anxiety (Stagl 1995:52). By contrast, tourism today is comfortable and relatively cheap. With the commoditization of tourism in modernity, leading to the transformation of travel into tourism, there is nostalgia for the lost art of travel (Boorstin 1964). The traveler and the tourist are treated as two different categories. Fussell has summarized the distinction between the explorer, the traveler and the tourist:

> All three make journeys; but the explorer seeks the undiscovered, the traveller that which has been discovered by mind working in history, the tourist that which has been discovered by entrepreneurship and prepared for him by the arts of mass publicity. The genuine traveller is, or used to be, in the

middle between the two extremes. If the explorer moves toward the risks of the formless and the unknown, the tourist moves toward the security of pure cliché. It is between these two poles that the traveler mediates (1980:39).

Fussell favors the explorer most and the tourist least, and the traveler is put in between. He denigrates tourism because it leads to the emergence of "pseudo-places", or what Boorstin (1964) calls "pseudo-events". Clearly, the nature of travel experiences has been transformed. Travel has been increasingly deprived of its spontaneity and independence and has been converted into tourism, i.e., "another routinized, packaged, commodity" (Watson and Kopachevsky 1994:645). Modern forms of transport have replaced original means of travel. As Böröcz (1996:50) observes, the relationship between modern tourism and travel is like that between printing, especially mass-produced book printing, and the manuscript. Moreover, with the increased ease and reduced cost of travel, tourism has gradually been disconnected from utilitarian functions such as education and turned into a form of leisure, fun, entertainment, and consumer culture. In late modernity tourism has also been democratized as a mass phenomenon within popular culture.

Several factors help bring about the transformation of travel, including advances in transportation technology, improved living standards, and associated increases in disposable income and leisure time. However, the transformation of travel into tourism, especially mass tourism, can be better understood within the context of industrial capitalism. Böröcz elaborates as follows:

> the emergence of mass tourism is organically connected to the spread of industrial capitalism so that the production of tourists, hosts, and the commercial relationship between them, that is, the tourism industry, is a logical extension of the general principle of industrial capitalism to the realm of leisure (1996:50).

The logical link between tourism and industrial capitalism is supported by a strong positive association between the emerging, uneven structures of leisure migration (tourism) and those of industrial capitalism (1996:48). Thus, the more a country's industrial base is developed, the greater its potential for tourism. For instance "England and west-central Europe enjoyed an advantaged, forerunner position, and the eastern portion of the continent suffered a consistently disadvantaged, latecomer status" (1996:45). Part of the reason for this is that, on the one hand, industrial capitalism created a need for leisure travel, and on the other, industrial capitalism developed a tourism industry to satisfy this need in a standardized, normalized, and commercial way (1996:45). Advanced industrial countries also established the necessary infrastructure for tourism development, and this is why these countries are the most toured destinations. Accordingly, Böröcz reconceptualizes tourism as "travel-capitalism" and claims that it is this contextual aspect that distinguishes tourism from various precapitalist types of travel (1996:29,50).

A crucial factor in the transformation of travel into tourism is capitalist commoditization. Tourism is in reality an extension of the essential logic of capitalism, namely the commodification of modern social life (Watson and Kopachevsky 1994:644, 645). In essence, "tourism is a form of the commoditization of experience" (Graburn 1983a:27). Commoditization has a number of characteristics. First, there is universalization. Commoditization discriminates against nothing, and in principle there is no single item that cannot be commoditized. Thus, commoditization involves the transformation not only of tangible objects, but also intangible services, activities, and experiences, into commodities (Watson and Kopachevsky 1994:649). In Fairclough's words:

> The *commodity* has expanded from being a tangible "good" to include all sorts of intangibles: educational courses, holidays, health insurance, and funerals are bought and sold on the open market in "packages", rather like soap powders (1989:35).

Under capitalism, commoditization is ubiquitous and almost anything can be turned into a market commodity. Tourism is no exception. In fact, it is the result of the commoditization of travel experiences. Tourism is a specific kind of experiential commodity.

Second, commoditization is rationalization. It is informed by instrumental rationality. It maximizes output with minimized input, or to put it another way, if a means (e.g., action) is most effective in achieving a given end, it is rationally chosen as that means. Ideal-typically speaking, under capitalism profit is the only goal of commodity production. Therefore, commoditization is centered on this axis. In order to maximize profit the cost of production must accordingly be minimized. In so doing, both products and the production process must be scientifically and rationally designed, and commodity production becomes standardized and routinized. Related to this general economic rationalization is the more specific "rationalization of the holiday product" (Swinglehurst 1974:200), such as Thomas Cook's commercial organization of tours. To put it more succinctly, the rationalization of the holiday product takes the form of packaging various disparate elements into a totality. This is what Thomas Cook and his contemporaries invented. Buzard sums up the situation as follows:

> More than did their numerous competitors, Baedeker, Murray, and Cook came to embody the power of rational administration over the many disparate elements that come into play in tourism—railways, custom houses, inns and hotels, currency exchange, regulations, and so forth, not to mention the diversity of interests and temperaments among the clientele they served (1993:48).

Such a rational integration of disparate elements into a total experiential commodity is, in fact, a form of control over the touristic experience, which leads to the transformation of travel into tourism (1996:51). As Rojek puts it, "In tourism, escape experience is packaged in an intensely commodified form" (1997:58). In the past travel experiences were often characterized by uncertainty, risk, and lack of predictability. Today, in order to qualify as a commodity that consumers will buy and consume, the product of tourism requires that uncertainty, risk, inconsistency, and contingency be reduced, or if this is impossible, quantitatively calculable and compensated by insurance companies. In such a manner, certainty, safety, consistency, and predictability are increased. Rationalization of the product of tourism has become indispensable.

Third, commoditization is standardization. Irrespective of whether the commodities are produced for the mass market or for specific segmented markets, the quality of the same kind of commodities must remain similar and consistent. Standardization is therefore a necessary measure in commodity production and includes the standardization of service quality, products, and experiences. Correspondingly, consumers evaluate the quality of holiday products in terms of certain objective criteria, such as whether the quality of the holiday matches the price charged (i.e., value for money). Such criteria are embodied in consumers' anticipations and expectations. If the quality of the product is greater than these expectations, satisfaction results. By contrast, if the quality of the product is lower than expected, then dissatisfaction will occur. However, standardization may lead to the homogenization and substitutability of certain destinations, as illustrated by the increasing similarity between beach holiday destinations.

Finally, commoditization is quantification. Although experiential products have an intangible form, they must be quantified in order to control costs and maximize profits. The product of touristic experiences usually takes the form of *itineraries*, since an itinerary can be quantified in terms of money and time. Tour operators can calculate in detail the cost of each constituent element of the product, such as the costs of advertising, tour operations, transportation, accommodation, entry fees to attractions, and insurance. Tour operators also bargain with destinations, and then wholesale the product to travel agencies (although some tour operators invest in destinations and have their own accommodation in those places). It is due to the principle of quantification that scientific management of the touristic experience becomes possible.

Through commoditization, touristic experiences are packaged as a saleable commodity. However, commoditization also leads to the truncation and insulation of experiences, a consequence that Boorstin (1964) criticizes as "pseudo-events". The raw materials of experiences are "cooked" as consumer "foods" (i.e., tourism) that are presented in the menus of tourism products (e.g., brochures). Thus, tourism becomes a consumption activity, radically different from travel in the past.

However, commoditization may also compensate for the deficiencies caused by standardization by diversifying the range of experiences on offer. New experiences are constantly invented and new itineraries organized. For example parachuting and submarine trips have been added to the "menus" of many tourist consumption sites (destinations). Thanks to commoditization, once the old products are no longer interesting, many more new products can be expected to be introduced. Indeed when the technology of space transportation becomes more advanced and feasible, then the commoditization of moon exploration should follow soon after. Moreover, as commoditization does not discriminate against anything, whatever tourists wish to consume in tourism sites may in principle be supplied. It is thus no small wonder that sex is heavily commoditized in Southeast Asia, a point that will be developed later on in this chapter.

Tourism as Experience-Centered Consumer Culture

In *The World of Goods*, Douglas and Isherwood (1996:132–133) define consumption in relation to the consumption of three sets of goods: a staple set corresponding to the primary production sector (e.g., food); a technology set corresponding to the secondary production sector (e.g., travel and consumer's capital equipment); and an information set corresponding to tertiary production (e.g., information goods, education, arts, cultural and leisure pursuits). Tourism is also a consumer activity, involving the consumption of technology (such as transport), signs and symbols (Watson and Kopachevsky 1994:650), services (such as accommodation), goods (such as souvenirs and films), and pleasurable experiences (Campbell 1995:118–119). Thus, the tourist is in nature a consumer rather than a producer (Costa 1993; Prentice 1997:209; Saram 1983:91; Vukonić 1996; Watson and Kopachevsky 1994). With the arrival of the democratization of leisure travel in the post-war period, Western society witnessed the integration of "touristic consumer culture" into consumer culture in general. However, there is an important difference between touristic consumer culture and goods-centered consumer culture. The former is characterized by a cultural orientation towards the consumption of *experiences* as a meaningful activity, together with the associated images and feelings attached to such consumption. By contrast, goods-centered consumer culture is characterized by a cultural orientation towards the consumption of *goods* and their associated images, meanings and feelings.

Touristic consumer culture is the extension of a goods-centered consumer culture to the sphere of "experiences", including music, spectator sports, films, and tourism. In order to distinguish touristic consumer culture from a goods-centered consumer culture, the former can be defined as an "experience-centered consumer culture". With non-material goods, such as information, advice, and expertise, leisure activities play an ever greater role in the postindustrial economy and consumption; they are turned into saleable commodities for consumption (Campbell 1995:110; Slater 1997:193). This situation leads to the *dematerialization* of commodity forms or consumer goods (Lee 1993:135; Slater 1997:194). Touristic experiences constitute one such example. Whereas the consumption of material goods takes place *during* leisure time, the consumption of experiences (or experiential commodities) is in fact the consumption of a certain period of time *itself*, namely the consumption of "a good time". Thus, time is what the experience-centered consumer culture is based on. As Lee puts it:

> Effectively, there has been a marked "de-materialization" of the commodity-form where the act of exchange centers upon those commodities which are time rather than substance based. The push to accelerate commodity and value turnovers has triggered a transition from the characteristic durable and material commodities of Fordism (washing machines, vacuum cleaners, automobiles, television sets, etc.) which have a lifespan extending well beyond the actual moment of exchange, to non-durable and, in particular, experiential commodities which are either used up during the act of consumption or, alternatively, based upon the consumption of a given period of time as opposed to a material artifact (1993:135).

Tourism is one of the most popular types of experience-centered consumption in late modernity. As a "sacred" period of time which is separated from a profane period of time (Graburn 1989), tourism is culturally highly valued and desired. Moreover, the popularity of touristic consumer culture is related to the expanding "experiential reach" in modernity. The constant transcendence of this experiential reach across national boundaries has accompanied advances in transportation technology and the associated commoditization of experiences. This kind of experiential commodity is organized as a commodity of itineraries and sold to customers for consumption during their travels. Thus, leisure travel, movement, and mobility have been incorporated into consumer identity under late modernity.

The Seduction of Tourist Space as Consumption Sites

As Langman comments:

> If the Gothic Cathedral was the symbolic structure of the feudal era, and the factory of the industrial, the distinct structures of today are cultural sites or

theme parks like the Centre Georges Pompidou of Paris or Disneyland, and the carnivals of consumption—the shopping malls (1992:41–42)

Indeed, what characterizes the current era is not production but consumption, especially experiential consumption. It can be argued that the consumption site is the symbolic structure of the postindustrial age, and tourist space is an increasingly significant consumption site.

Loosely speaking, under modernity there is a gradual differentiation between production site and consumption site (of course it cannot be denied that production also involves the consumption of materials, but here consumption refers not to productive consumption but to "end consumption"). For example, in premodern society the home was both a unit of production and a unit of consumption, whereas in modern society it is exclusively a site for consumption. For modern people, much of their consumption is conducted at home or is centered upon the home. However, there are also consumption sites outside home, such as bars, restaurants, theatres, cinemas, and shops, which can be referred to as public consumption sites. Analytically speaking, the formation of these public consumption sites has to do with the process of urbanization. In the age of industrialization the city was not only a principal site of production but also a major market, that is, a site of consumption. However, with the arrival of post-industrialization in the West the city has been transformed "from the industrial city to the city as a place of consumption, entertainment and services" (Slater 1997:202). Postindustrial cities have increasingly been equipped with consumption, leisure, and entertainment facilities in order to attract tourists, other international movements of people, and those middle-class consumers who moved to the suburbs during late modernity (1997:202). As a result, the city as a whole has become a major public consumption site.

In addition to the home and the city, a third major consumption site is the touristic space, such as holiday resorts and other destinations. Cities, including global cities such as New York, Hong Kong, and Tokyo, and cultural and historical cities such as Rome, Paris, London, and Beijing, are major tourist destinations, and hence touristic consumption spaces. Many destinations are within Third World countries. Thus, the globalization of tourism results in the global extension of consumption sites for tourists from economically rich countries.

Unlike commodities that can be consumed at home, tourism must be consumed outside the everyday social space (Thurot and Thurot 1983:175). Although consumption at home and outside the home in a tourist space may both bring pleasure, consumption in the latter is conducive to certain illusions and feelings that do not apply to consumption at home. Tourist space, which is often far away from home, becomes a symbol of the freedom of spending, freedom from the constraints of spending at home. As Watson and Kopachevsky note:

> for many touristic experiences, a "great vacation" is going to a place where opportunities to spend money and watch others perform, dominate their time. A good time, it is suggested, has come to be associated with spending and consumption, and pleasure may be measured by spending (1994:650).

Being free from the pressure to earn money in the home society, tourists can enjoy the fantastic feeling of non-constrained spending. For many, this is a reward for the monotony of their daily work. However, the momentary experience of free spending often involves a sum of money that has been saved over the course of a tiresome working year. Nevertheless, spending on holiday can become a *consumption ritual* that offers people pleasure and fantasy. With the seduction of such hedonistic activity, the daily malaise of work in the home environment may be made more tolerable.

From a supply-side perspective, consumer culture is related to the services provided by both the tourism industry and touristic destinations. The industry is a major agent of consumer services, i.e., the production of "experiences as commodities", typically in the form of itineraries. Touristic destinations are major (collective) actors in the production of tourism. As noted above, under Western modernity the supply of tourism services is the extension of the logic of capitalist commodity production to the sphere of leisure and tourism, that is, the commoditization of itineraries and their associated experiences. This leads to a consumer culture that is characterized by "imaginative pleasure-seeking".

Tourist consumer power has obvious economic effects upon local economic development. For example, in the postindustrial age, to alleviate the bitterness of structural unemployment caused by the decline in traditional manufacturing, various postindustrial cities in Northern England, such as Sheffield, Bradford, and Manchester, have created an image aimed at attracting visitors to "consume" their places (Bramwell and Rawding 1996; Roche 1992, 1994). This response to the opportunity brought about by touristic consumerism has been intensified. Indeed, various places are vying for domestic and international tourists in order to stimulate the local economy and increase employment opportunities.

To take full advantage of touristic consumerism, a place must exploit and commodify the locally available resources. Thus, thanks to tourism, some "ordinary" and "taken-for-granted" things are turned into attractions, such as vernacular houses in Beijing (Wang 1997a). In places where there are no natural or cultural attractions, the construction of artificial attractions such as theme parks, shopping malls, and conference facilities can be undertaken in order to increase their "pulling" power. A typical example in this respect is Shenzhen (a Chinese city near Hong Kong), a place that was set up as a special economic zone in 1980. At the time, Shenzhen was only a village, but it has now become a large city with millions of inhabitants. It has also become one of the most popular tourist destinations in China, particularly for tourists from Hong Kong. Once characterized by a lack of both natural and cultural attractions, Shenzhen's success has relied upon heavy investment in the construction of touristic attractions, including theme parks such as The Window of the World, The Village of Chinese Customs and Splendid China, all of which have been highly economically successful. As a result Shenzhen, a newly industrialized city, has become a "pleasure periphery" or "leisure backgarden" of Hong Kong.

It can be argued that contemporary society has witnessed a process of *touristification*, a socioeconomic and sociocultural process by which society and its environment have been turned into spectacles, attractions, playgrounds, and consumption sites. The existence of tourist information offices in various places across Europe, Australasia, and North America is in fact an index of the extent of touristification in the West.

As discussed earlier, according to Böröcz (1996), touristification is uneven because economic development is uneven. The more advanced an economy, the more advanced its touristification. Developed countries have several advantages when touristifying themselves. First, they have adequate infrastructure and facilities. Second, they have advanced transportation systems and networks. Third, they have a relatively sufficient supply of travelers due to the increasing flow of people between places for economic and other reasons, which in turn leads to investment in tourism-related facilities, such as hotels, restaurants, bars, and entertainment sites. Fourth, they have sophisticated marketing techniques which can help enhance their image. In addition, the West-centered discourse that circulates within the world media also raises the reputation of certain destinations in developed countries. Fifth, the economic power of advanced nations affords them greater opportunities to host world mega-events such as the Olympic Games and world conferences, which attract the attention of a global audience and consequently increase place-reputation and symbolic capital, which are closely bound up with and can be turned into economic capital.

However, touristification also occurs in developing countries. Many nations in Africa, Asia, and Latin America have joined the world competition for tourists. They have become integrated into a network of global consumption sites for international tourists from economically rich countries. Touristification is therefore a universal phenomenon. It is an integral element of globalization. Thanks to global touristification, the flows and circulation of people, capital, money, images, and commodities are becoming rapidly interlinked and intensified. It is also true that consumerism, which originated in advanced countries, has spread to developing countries, where the poverty of the inhabitants stands in sharp contrast to the hedonistic consumerism of outsiders.

However, as much of the literature has documented, touristification in developing countries is handicapped by the initial lack of adequate infrastructure and facilities. Such a situation leads to the so-called the problem of "dependency", since many developing countries rely upon foreign capital to construct the infrastructure and facilities which are necessary for touristic consumption. Touristification in developing countries is also influenced by image fragility and the uncertainty of attracting foreign consumers. It has been said that a cough in the tourist-sending market will bring about a cold in some Third World countries, where the contribution of tourism to the national economy is quite significant. Touristification in developing countries is thus much more problematic than in advanced countries (Turner and Ash 1975).

In some developing countries, especially some Southeast Asian countries, touristification can lead to the eroticization of a place. Thus, sex tourism becomes an attraction in places such as Thailand, the Philippines, and South Korea (Hall 1992, 1994; W. Lee 1991; Leheny 1995). The consumption of exotic sex and the perceived submissiveness and docility of Southeast Asian women can indeed represent a magic and hedonistic experience for many males from economically advanced countries such as Japan and the United States (W. Lee 1991). The governments of some Southeast Asian countries are ambivalent about sex tourism. On the one hand, there is the fear of a bad moral image and the spread of diseases such as AIDS. On the other hand, there seems to be a large tolerance of prostitution and both overt and covert encouragement of sex tourism as a source of foreign exchange (Hall 1994:143). The reasons for such governmental encouragement are complex. One reason is cultural, for prostitution is an accepted tradition in certain countries. Another reason is the unequal power balance between economically rich and poor countries, which in turn leads to an unequal global division of labor in tourism and indecent touristic consumption, i.e., the consumption of sexual services more often than not takes place in developing countries. While not denying that sex tourism also exists in some advanced countries, such as Holland, the social consequences of sex tourism in developing countries are much more serious. For example, child prostitution and the spread of AIDS have become serious issues in some southeast Asian countries. As exotic sexual attractions characterizing the tourist consumption sites of Southeast Asia, sex tourism is exploited in these countries as a means towards economic development. In some of these countries, any kind of activity that can bring in foreign currency is valued. Almost anything is tradable and can be touristified, including women and even female children's bodies. Thus, with the globalization of touristic consumerism, exotic sex becomes a major seductive "dish" on the "menu" offered by consumption sites. Despite the realization that such a "dish" may be spoiled by the spread of Aids, it has survived and will continue for a long time. For many males from economically stronger countries the prostitutes of Southeast Asia are adventurous, exciting, and exotic attractions. The touristification of Third World is thus a highly ambivalent process.

In summary, for tourists, tourism is essentially a consumer activity. This involves the commoditization of tourism, which has transformed the travel of the past into contemporary tourism as a consumer good. However, unlike tangible consumer goods, tourism is an intangible commodity of "experiences", exhibited in the form of itineraries. Under late modernity there emerges a touristic consumer culture which is distinguishable from a goods-centered consumer culture. The former is experience-centered, and one of its essential features is a hedonistic orientation towards the consumption of time itself. Experiences are thus regarded as the essence of "a good time", time to be consumed. With the advent of this consumer culture, various places strive for the touristification of their cultures, heritage, and environments, which form part of the postindustrial economic strategy in advanced countries. With the globalization of touristic consumer culture, namely the global extension of the reach of tourism, many Third World countries join in with the touristification of their cultures, people, and environments. However, due to a relatively weak position in international competition, a "dehumanizing" form of tourism, i.e., sex tourism, particularly child prostitution, appears in a number of Third World countries, such as some in Southeast Asia. This type of tourism is implicitly accepted, sometimes even encouraged, explicitly or implicitly. Thus, the globalization of touristic consumer culture may arguably bring about problematic social consequences, particularly for some Third World countries.

Chapter 10

The Lure of Sign Value

As discussed earlier, tourism is a consumer activity which consists of both experiential and symbolic consumption. The former was examined in the previous chapter. The latter is the subject of this chapter. From a longitudinal perspective, tourism as consumption consists of three consecutive stages: pre-travel, on-trip experiences, and post-travel. The pre-travel stage involves the consumption of images and representations, and has been touched on in *chapters 7* and *8*. On-trip experience involves the consumption of experiences, or experiential consumption, and this phase has been discussed in *chapter 9*. The post-travel stage involves the symbolic effect of the consumption of tourism, which has not yet been explored. Thus, here the focus will be on this stage, one which is exemplified by the post-travel practice of bringing home souvenirs. This chapter consists of three sections. The first conducts a review of literature on the symbolic character of consumption in general. The second, accordingly, treats tourism as symbolic consumption. The third analyzes how souvenirs function as symbolic consumption.

A Brief Review of the Literature on Symbolic Consumption

The idea that consumption has symbolic significance is not new. Veblen (1925) proposed this thesis in his classic work *The Theory of the Leisure Class*, originally published in 1899. He coined the term "conspicuous consumption" to define the characteristics and styles of the leisure class's consumption. Rather than consuming goods and services for their functional utility, Veblen argues that the leisure class's consumption is for show or conspicuous display alone, thereby serving to enhance their social status. This idea is reexamined by Bourdieu in his influential *Distinction*. For Bourdieu (1984), consumption is of symbolic significance because it is a signifying practice, parading the taste and lifestyle that are associated with social locations. Consumption is thus symbolic largely because of what the consumer consumes. What is consumed is dictated by the cultural codes of each social class, which underpin the classification and selection of goods for consumption (Bourdieu 1984). Hirsch (1977) also treats consumables as "positional goods".

In *The World of Goods*, Douglas and Isherwood argue that "We can never explain demand by looking only at the physical properties of goods. Man needs goods for communicating with others and for making sense of what is going on around him" (1996:67). Thus, echoing Veblen, they suggest that goods are not consumed solely for their physical use, but are also "coded for communication", serving to "make and maintain social relationships" (1996:xxi,39).

Baudrilland deals with the symbolic aspect of consumer culture from a semiotic perspective. He argues that a commodity has a third value, i.e., a sign value, in addition to what Marx has termed use value and exchange value (Baudrilland 1988:57-97). "In order to become object of consumption, the object must become sign" (1988:22, emphasis deleted). Thus, for example,

> a washing machine serves as equipment and plays as an element of comfort, or of prestige, etc. It is the field of play that is specifically the field of consumption. Here all sorts of objects can be substituted for the washing machine as a signifying element. In the logic of signs, as in the logic of symbols, objects are no longer tied to a function or to a defined need. This is precisely because objects respond to something different, either to a social logic, or to a logic of desire, where they serve as a fluid and unconscious field of signification (1988:44).

Hence for Baudrillard, "consumption is neither a material practice, nor a phenomenology of 'affluence'" (1988:21), but rather "*a systematic act of the manipulation of signs*" (1988:22). Consumption is therefore a socially meaningful and culturally signifying semiotic practice.

Because of the symbolic function of consumption it is not surprising that people use it as a means of demonstrating identity. Williamson asserts that, "instead of being identified by what they produce, people are made to identify themselves with what they consume" (1982:13). Consumption is thus a solution to the problems of identity in contemporary Western society (Slater 1997:85). "We are made to feel that we can rise or fall in society through what we are able to buy, and this obscures the actual class basis which still underlies social position" (Williamson 1982:13).

Consumption is linked to the reproduction and obtainment of status. Of status, Csikszentmihalyi and Rochberg-Halton write:

> Status is also a form of power, but of a different kind from the raw kinetic energy contained in spears and cars. It consists of the respect, consideration,

and envy of others. A person with status sets the standards and norms by which others will act, and in this way embodies the goals of a culture. Similarly, a thing with status also acts as a template embodying these goals because it will cause people who believe in its status to act accordingly toward it and its owner who possesses the status (1981:29).

Such being the case, things that have the power of "status-giving" (1981:29) and act as status symbols later become objects of desire, display, and consumption. As Slater puts it, "goods, by virtue of their meanings, are tools of social climbing, social membership and social exclusion—their basic nature is to differentiate, but solely with respect to social hierarchy" (1997:153). Consumption naturally becomes a domain of status competition and acts as the cultural organization of social order by enabling visible social divisions, categories, and ranks (1997:150).

Tourism as Symbolic Consumption

Like other types of consumption, touristic consumption is of symbolic significance. It involves the consumption of not only substantial experiences which are packaged as commodities, but also the symbolic signs that demarcate social status and group identity (Brown 1992; Thurot and Thurot 1983). Touristic consumption thus involves symbolic capital (Bourdieu 1984; Munt 1994; Urry 1990a). Thurot and Thurot (1983) apply Baudrillard's idea of the sign value of a commodity to tourism and reveal the symbolic aspect of tourism through an examination of advertisements. Moeran (1983) describes how Japanese tourists are preoccupied with the status derived from touristic consumption abroad. Based on a review of the literature, Brown discusses the relationship between tourism and symbolic consumption, and concludes that tourism can be portrayed as "a multi-faceted symbolic act" (1992:58). It is no wonder that a suntan acquired on holiday is portrayed as a status symbol (Reimer 1990:508). Indeed, holidays are themselves status symbols (Turner and Ash 1975:284).

Dimanche and Samdahl (1994) have examined the symbolic significance of leisure consumption, which is also applicable to touristic consumption. For them, leisure consumption is "primarily symbolic and reflects a subjective state of consciousness that is heavily dependent on symbolic meanings, hedonic responses, and aesthetic criteria" (1994:122). The symbolic consumption of leisure has two aspects: self-expression and sign value. Self-expression, as an expression of personal identity and communication to self, requires an "internalized social audience" or a "private audience'. By contrast sign value, as an expression of social identity and communication to others, entails "an external audience"; "leisure consumption needs to be conspicuous because the important attribution of traits comes only from the reaction of others through the visible nature of consumption" (1994:124).

In a similar vein, Prentice claims that touristic consumption is "a mix of internalized meanings and external symbolism" (1997:210). Thus, tourism "can only be understood fully in terms of what the consumer is seeking to get out of their visit directly, and indirectly through the reactions of others in their home community as symbolic additional value" (1997:210–211). Costa (1993) examines touristic consumption as identity expression from an alternative angle: for Western Europeans, tourism is consumed in order to display their uniqueness and differentiation from their hosts, especially hosts in Eastern European countries such as Hungary.

However, if the experiential aspect of touristic consumption is linked to hedonism, then the symbolic aspect of consumption can lead to social pressure. With the institutionalization of holidays with pay and the consequent democratization of tourism, "not to travel" may become an indicator of deprivation or failure, namely a signifier of a person's state of health (too old or too sick to travel) or failure in career (hence unable to afford to travel), because good health and success are widely accepted values in Western society. Thus, a healthy or a successful person may feel obliged to take at least one holiday a year in order to avoid the social stigma of failure and gain tacit or verbal social acknowledgement, respect, and even the envy of other members of the home community. Consequently, as Vanhove claims,

> For most Europeans a holiday each year is a "must", even when the economic situation is temporarily less brilliant. Economic crisis may mean the destination is different or the stay shorter, but none the less the participation rate is more or less unaffected (1997:56).

Taking a holiday is thus a matter of personal symbolic standing in society, and not merely functional in terms of physical rest and psycho-physiological refreshment. In this sense, consumption involves "status consumption" (Gilbert 1994:123). Wright Mills (1951) characterizes the latter as the "status cycle". As he notes, there may

> be a more dramatic yearly status cycle, involving the vacation as its high point. Urban masses look forward to vacation not "just for the change", and not only for a "rest from work"—the meaning behind such phrases is often a lift in successful status claims. For on vacation, one can *buy* the feeling, even if only for a short time, of higher status. The expensive resort, where one is not known, the swank hotel, even if for three days and nights, the cruise first class—for a week. Much vacation apparatus is geared to these status cycles; the staffs as well as clientele play-act the whole set-up as if mutually consenting to be part of the successful illusion. For such experiences once a year, sacrifices are often made in long stretches of grey week-

days. The bright two weeks feed the dream life of the dull pull (1951: 257–258).

Therefore, vacations or status cycles "provide, for brief periods of time, a holiday image of self, which contrasts sharply with the self-image of every-day reality" (1951:258). Touristic consumption becomes a social tool to enhance self-image and compete for status and cultural and symbolic capital. To put it another way, tourism is not only an accumulation of *experiences*, but also an accumulation of *reputation, status, and symbolic capital*. Thus, the symbolic consumption of tourism becomes a strategy in striving for symbolic power in society.

As previously mentioned, taking or not taking a holiday may affect a person's status. However, with almost everybody in the West engaging in tourism, mere participation in a holiday is not a sufficient condition for gaining status. Status is also related to *where* the holiday is taken and to the type of touristic consumption. For example, in the early stages of tourism, seaside resorts in England were almost solely occupied by the upper and middle classes because of their economic power. However, with the arrival of mass tourism in these resorts, the upper and middle classes tended to avoid these areas and went instead to more remote and expensive destinations abroad. This served to maintain their status and prestige because the lower class were excluded.

However, some members of the lower class may compete for higher symbolic status in cultural rather than financial terms. As Munt (1994) points out, although the new petite bourgeoisie cannot afford expensive and luxurious holidays because of a shortage of *economic* capital, they may adopt alternative styles of holidaying in order to maintain and enhance their *cultural* capital. They tend to take a holiday "with an obsessional quest for the authentication of experience", usually in Third World des-tinations with a "true and real contact with indigenous peoples" (1994:108,111). Such a new style of holiday-making or travel (so-called alternative tourism) helps compensate for their shortage of economic capital, and hence they can claim cultural sensitivity and superiority. Tourism is therefore stylized. As Graburn puts it,

> changes in tourist styles are not random, but are connected to class competi-tion, prestige hierarchies, and the succession of changing life styles, as well as to external factors such as the cost and modes of transportation, access to regions and countries, and the state of the economy. Each style is chosen (a) with reference to the previous experiences of the individual (and group or class), such as family and friend's experiences, and (b) with reference to competing patterns of other classes or groups (or individuals) whom one wishes to emulate or to avoid. Thus … tourist style never stand still (1983a:24).

The choice of tourist style is influenced by several factors, two of which are relevant here: the first is the reference groups with which individuals would like to identify; the second is status competition. No matter which style is chosen, touristic consumption is stylized both for self-expression and in competition for symbolic capital (status).

The Sign Values of Souvenirs

As mentioned earlier, both tourism in general and tourist styles in particular act as signs. However, since touristic consumption is both a private affair and an intangible experience, it may often be unknown to a person's social interactants or community at home. Therefore, symbolic consumption and conspicuous display requires visible and tangible evidence to impress one's community, or remind oneself that one has gone somewhere for a certain kind of holiday. As Csikzentimihalyi and Rochberg-Halton state: "It is difficult to imagine a king without a throne, a judge without a bench, or a distinguished professor without a chair" (1981:15). Similarly it is also difficult to imagine a tourist without a camera or souvenirs. If a camera is the tourist's primary "identity badge" (Chalfen 1979:436), then souvenirs, together with photographs of the destination, are convincing evidence of the tourist's trip. The consumption of souvenirs is thus an integral aspect of symbolic consumption in tourism (Brown 1992; Costa 1993; Gordon 1986; Littrell 1990; Littrell et al 1993, 1994; Moeran 1983; Voase 1995).

Indeed, as Voase (1995) points out, from a pragmatic perspective souvenirs hardly have any functional value within the range of regular consumption. Nor can they be regarded as household essentials. In reality, souvenirs are often viewed by both purchaser and recipient as "something of a nuisance" (1995:48). Therefore, the actual function of the souvenir, and hence the reason for its purchase, must lie elsewhere in terms of its symbolic and signifying properties. Hence, "the stick of rock symbolizing the seaside, the Eiffel Tower symbolizing Paris, documenting the fact that the visit took place" (1995:48). The purchase, acquisition, or giving of souvenirs contains the "secrecy" of the symbolic consumption of tourism in general, a topic which deserves detailed exploration and elaboration.

Souvenirs as Signs

As mentioned above, the temporal and longitudinal structure of touristic consumption consists of three consecutive stages: the pre-travel imaginative experience (i.e., imagination, dream-chasing, and anticipation *before* the journey), the ongoing substantial experience (experiential consumption *during* the journey), and the post-travel experience of remembrance (reminiscence of holiday experiences *after* the journey). All these phases are integral to touristic consumption. At the pre-travel stage images are consumed. At the stage of ongoing travel, substantial experiences linked to itineraries and services are consumed. At the post-travel stage it is impressions and memories as well as symbolic objects, such as photographs and souvenirs, that are consumed. Moreover, this last stage assumes the major function of conspicuous display. The purchase and consumption of souvenirs thus gain symbolic significance because they serve as visible proof of the journey, which may otherwise remain unknown to the interactive groups and community at home.

What kind of things can become souvenirs? Gordon (1986:139–144) classifies souvenirs into five types. First, pictorial images, such as the postcard. Second, 'piece-of-the-rock' souvenirs: these are parts of a whole, usually natural materials or objects. Third, symbolic shorthand souvenirs, usually manufactured rather than natural objects. Fourth, markers such as T-shirts printed with words relating to the destination. Finally, local product souvenirs such as indigenous foods, liquor, identifiable local clothing, and local crafts.

Dimanche and Samdahl's (1994) classification of two aspects of the symbolic consumption of leisure can be applied to the symbolic functions of souvenirs, namely internal communication and external communication. Accordingly, souvenirs can act as both outward signs and inward signs. By "outward sign" is meant that the meanings of souvenirs and of souvenir-giving are culturally and socially coded and decoded, so that the souvenir functions as a show, a conspicuous display, and a social communicator, conveying certain messages and meanings to a social (external) audience. In this sense souvenirs can be both a sign of social re-entry or social acceptance and a sign (or symbol) of status. By contrast, an "inward sign" means that the meanings of souvenirs are personally attached. The purchase and consumption of souvenirs are, though not exclusively, for personal taste, a hobby, and the symbolic representation of a travel career, which communicates self-identity to an internal audience (internalized social audience). In this sense, souvenirs can be both a sign of a meaningful biographical event and a sign of self-identity. It should be noted that the distinction between "outward sign" and "inward sign" is *analytical*. In reality, they often coincide with one another.

Souvenirs as Post-Holiday Gifts and Signs of Re-entry

In some cultures, tourists returning home are required to offer gifts to their friends, relatives, or colleagues as a re-entry "fee" to, or part of the rite of reincorporation into, their culture (Gordon 1986; Moeran 1983). Souvenirs thus serve as "post-holiday gifts", and the giving of these "has much to do with social *ritual*, of which the need for reciprocation is a key feature" (Voase 1995:48). For example, the long existing gift culture in Japan requires tourists abroad to purchase gifts for their friends and relatives at home (Moeran 1983). As a result, one of the outstanding features of Japanese tourism brochures is detailed information on the opportunities for buying souvenirs and detailed descriptions of the kinds of souvenir available in various destinations. As Moeran discovered from an analysis of such brochures:

> Shopping abroad, therefore, is emphasized in the brochures because it gives the traveler some indication of the sort of gift he can—indeed must—bring home and present to family, friends, and fellow workers (1983:99).

Thus, souvenirs play an important role in the ritual of re-entry into the home society after a temporary absence. The giving of souvenirs is a symbolic demonstration of the resumption of their previous social roles and connections to their network of social relationships at home. Based on such a ritual of desacralization, which indicates the end of the ritual of "sacred passage" (Graburn 1989), returning tourists are socially "welcomed" by their community. When the recipients of souvenirs express their thanks, inquire about the destination and the journey, the giving of souvenirs turns the holiday into news, a performance, a *socially acknowledged and sanctioned* biographical event, and consequently removes the possibility of returning home without social attention. The consumption of souvenirs is therefore rarely a private matter, but more often than not a socially significant act to which, according to Max Weber (1978), subjective or intersubjective meanings are attached. In short, souvenirs are part of the neo-folklore of symbolic exchanges in an emotional community.

Souvenirs as Status Symbols

Souvenirs are not only gifts presented to relatives and friends, but are also social communicators which display the status derived from a journey. Indeed since tourist experiences are *intangible*, it is necessary to convince others that the visit took place by means of visible and tangible evidence, such as souvenirs and photographs. Thus, souvenirs are *outward signs* that signify status: they are symbols of status. As Brown claims, tourist goods "can act as social tools when their symbolic qualities promote communication between an individual and his or her significant references" (1992:58). In Watson and Kopachevsky's words, souvenirs "symbolize the tourists' experience; when bought and taken home for self or friends, they are indeed tangible evidences of travel, a representation of a gaze or experience" (1994:652). Böröcz also points out that souvenirs, along with photographs, family travel films, and videos, "work as symbolic preservatives of travel experience and a means of communicating the same to nontraveling members of the home society" (1996:9). It is the communicative and symbolic functions of souvenirs that constitute the reason why tourists can hardly refuse to buy souvenirs when holidaying. Moeran's study of Japanese tourists' purchase of goods is a case in point. As he notes:

> one of the major preoccupations of the Japanese tourist is that of status...Thus one of the pleasures—and indeed for some, one of the "musts"—of going abroad is that of being able to purchase a brand name commodity in its country of origin...It is this kind of shopping item that the Japanese tourist keeps for himself, not just a reminder of a trip abroad, but as a symbol of status that comes from having been abroad (1983:99–100).

Turner and Ash speak of photography as "the technology which allows one to get the best of all worlds by turning tourism into a piece of selectively conspicuous consumption" (1975:283–284). The same observation applies to souvenirs. The symbolic consumption of tourism is closely related to the social symbolism of souvenirs, which helps enhance the tourist's public image. However, with most people being able to take holidays, the giving of souvenirs is symbolically differentiated in terms of where the souvenirs are produced and purchased, and whether they are authentic. Usually, the more exotic and remote the place where the souvenir is produced and purchased, the higher its symbolic status. Likewise, the more authentic the souvenir, the higher its status. Souvenirs thus act as differentiating signs of status.

The Souvenir as a Sign of a Meaningful Biographic Event

In addition, to acting as outward signs, souvenirs also function as inward signs. The act of souvenir purchasing is imbued with personal subjective meanings. First, souvenirs serve as signs of meaningful biographical events. From a biographical perspective, life history is constructed out of a series of significant and meaningful events, such as a birthday party, graduation from school, college, or university, the beginning of a career, marriage, promotion, and awards. Thus, biographical events are different from daily routines. They are rather the rupture of these routines and are significant to biographical identity. A whole life history is in fact "marked" by a number of life events rather than routines. If a year is taken as the basic unit of biography, then holiday-making is a meaningful biographical event within that year. It is an annual or semi-annual ritual of respite from the daily tempo and an escape from weekly routines.

The reason why a souvenir functions as a sign of a meaningful biographical event is that it serves as a "messenger of the extraordinary" and helps "bring back into ordinary experience something of the quality of an extraordinary experience" (Gordon 1986:140). To put it another way, "People like to be reminded of special moments and events, and a souvenir serves as such a reminder; indeed, the word itself means to 'remember'" (1986:135). The souvenir is an authentic reminder of a particular place or country which has been visited (Shenhav-Keller 1993). Being "a durable and portable signifier" and "an extension of the primary semiosis of the sight" (Harkin 1995:657), the souvenir is therefore purchased to memorialize the tourist experience as a series of extraordinary moments (Anderson and Littrell 1995; Gordon 1986; Graburn 1989; Hahn 1990; Jules-Rosette 1984; Littrell 1990; Littrell et al, 1993, 1994). As Gordon writes:

> People feel the need to bring things home with them from the sacred, extra-
> ordinary time or space, for home is equated with ordinary, mundane time
> and space. They can't hold on to the non-ordinary experience, for it is by
> nature ephemeral, but they can hold on to a tangible piece of it, an object
> that came from it (1986:136).

Since souvenirs serve as "tangible evidence of having found the authentic or having participated in the indigenous life of a community" (Littrell 1990:230), they commemorate sacred spaces and time (Costa 1993) and "extend the life of the tourism experience" (Brown 1992:57). Thus, for instance, a souvenir of the Great Wall of China helps memorialize a sacralized experience of China, and a souvenir of the Eiffel Tower serves to extend tourist life in Paris.

Extraordinary tourist experiences constitute authentic moments of an "experiential career". They are, therefore, regarded as meaningful biographical events. A biography is composed of an "experiential career" that consists of numerous memorable moments and life events. Functioning as reminders of the extraordinary moments of an experiential career, souvenirs, along with other symbolic objects, therefore play a certain role in constructing personal biography and self-identity.

The Souvenir as the Prop of Performance

The concept of self is central to the individual. According to Mead (1934), the self emerges from social interaction in which human beings internalize the attitudes of real and imagined others through "taking the role of the other". In this self-genesis process the "I", as the subjective aspect of the self, is involved in a continual interaction with the "me", i.e., the objective aspect of self (the object of self-consciousness). The "me" is the self as seen by others and represents the attitudes of the social group, the generalized other. Through role-taking in play and the "imaginative rehearsal" of interaction, individuals internalize the group's values as their own. By continually reflecting on themselves as others see them, they become skillful in the production and display of social symbols.

There are two kinds of social symbol: behavioral symbols and object symbols. Both are incorporated into the performance in which the self is presented to others, either internalized (imagined) others, or inter-active others. Souvenirs can thus be regarded as object symbols. In touristic consumption these are part of the communication of social identity: tourists perform using the props of souvenirs for the expected attitudes of imagined others. Appearing as trivial, souvenirs act as important symbols in constructing the rank in which people are placed and the class of taste to which they belong. Thus, as Costa points out, "the tourist purchases and displays souvenirs as part of the perfor-mance in which s/he presents self to others in his/her social domain. In the tourist's presentation of self, the purpose is to present self both as 'world traveler' and as one who appreciates the 'traditional' and 'authentic'" (1993:301). Littrell et al relatedly write: "Tourists use their souvenirs to reminisce, differentiate the self from or integrate with others, bolster feelings of confidence, express creativity, and enhance aesthetic pleasure" (1994:3).

However, tourists may bring home souvenirs but not give any of them to others. They may prefer to keep them without the purpose of display-ing them to the public. If such cases, the souvenir can still be a prop in a performance of the self, for the internal communication of the self is simultaneously external communication. The latter involves an actual social or external audience, while the former does not involve the pre-sence of others as a social audience. However, an internal communica-tion of the self involves "myself" as an audience, that is, an *internalized* social audience. Therefore, souvenirs are simultaneously both outward and inward symbols of the self.

More specifically, a souvenir is a means of expression of specific tastes. Tourists are not homogeneous (Cohen 1972). They do not con-stitute a single class. Rather, they are distinguished into different classes in terms of economic capital and cultural capital. As far as the latter is concerned, there are culturally sanctioned tastes in the consumption of various forms of tourism and touristic objects. The purchase of souvenirs does not merely convey the message that "I was there", and thus incorporate touristic consumption as part of a person's social identity. As mentioned above, many more messages are involved, including where the souvenirs were bought and whether the souvenirs are "authentic" (Littrell et al 1993). Hence, there are different kinds of touristic identity, just as there are different kinds of social identity in general, and souvenirs help to demonstrate one specific kind of touristic identity.

Cultural Trivialization versus Social Symbolism

Souvenirs constitute a symbolic world. However, this does not mean that there are no cultural misunderstandings between the givers (or holders) and the recipients (or viewers) of souvenirs. Therefore, the "consensual interpretation of symbolic meaning is a prerequisite for successful communication. It is, therefore, important to consider the means by which symbols attain a shared reality" (Brown (1992:59). A culture may have its own ways of coding and decoding the meanings of souvenirs, whereby people within the same culture can smoothly communicate with one another. However, the home society may "read" souvenirs bought abroad in terms of their own "ways of seeing" rather than in terms of the meanings and messages coded by the producers in the country of origin. If the latter treat souvenirs as symbols of national and ethnic identity (Shenhav-Keller 1993), then most tourists treat them as a sign of the exotic or ethnic, a sign that implies their presence at the site where the souvenirs are sold.

Thus, difficulties may arise in the cross-cultural communication between the producers of souvenirs and international consumers of souvenirs. From the former perspective the souvenir industry involves both the "commercialization of culture" (Turner and Ash 1975:140) and a "communicative act" (Jules-Rosette 1984:9) (for discussion of tourist arts or souvenirs from the perspective of the producer see Cohen 1993; Graburn 1976, 1984; Jules-Rosette 1984). On the one hand, souvenirs, as tourist art, are perceived as "a process of communication involving image creators who attempt to represent aspects of their own cultures" to tourists (Jules-Rosette 1984:1). On the other hand, however, the production of souvenirs is dictated by the principles of commercialization, where souvenirs are produced for an "external" audience (Graburn 1976). Therefore, local producers of souvenirs must produce what international tourists like and prefer to see, rather than genuine representations of authentic cultural traditions. This dilemma leads to the situation that Boorstin (1964) criticizes as a "pseudo-event" and MacCannell (1973, 1976) describes as "staged authenticity". "Tourist art mirrors the consumers' expectations and reveals the artist's perceptions of what consumers want" (Jules-Rosette 1984:3). In many cases, the authentic representation of local culture gives way to the commercial mass-production of souvenirs.

Thus souvenirs, especially those which are mass-produced, are frequently criticized as cultural trivialization, low in cultural meaning and artistic value.

Of course, it would be an overexaggeration to claim that commercialization necessarily leads to the destruction of the meanings of local and ethnic cultural products (Cohen 1988b). The relations between commercialization and ethnic arts (i.e., a kind of souvenir) are complicated and of various types. Erik Cohen (1993:3) identifies four types of commercialization of ethnic crafts as tourist art: complementary, substitutive, encroaching, and rehabilitative. Only the third type causes the degradation of local ethnic cultural traditions. In the case of rehabilitative commercialization, local ethnic crafts obtain opportunities for survival and rehabilitation.

Despite the fact that souvenirs are often labeled as a nuisance and kitsch, they are nevertheless quite popular and have become an increasingly important part of the tourism industry. Is the souvenir industry built on the degradation of culture? This issue still seems to be controversial. However, it is certain that the continuous survival and growth of the souvenir industry has much to do with the social symbolism of souvenirs (i.e., souvenirs as the symbols of status). Thus, many tourists may not consider whether souvenirs are authentic representations of local ethnic and cultural traditions in a strict sense when they buy them. The emphasis is placed upon the social symbolism of souvenirs which are "exotic".

Therefore, from the perspective of the consumer, many tourists purchase souvenirs not because they want to understand local people, history, and culture, but because the "exotic" souvenir is part of their conspicuous consumption. In this sense the production of souvenirs serves the hegemony of touristic consumerism rather than contributing to cross-cultural communication. Thus, as Stewart argues, the souvenir is "symptomatic of the more general cultural imperialism that is tourism's stock in trade. To have a souvenir of the exotic is to possess both a specimen and a trophy" (Stewart 1984, quoted in Frow 1991:145).

Souvenirs are thus imbued with social symbolism and their consumption provides an accumulation of symbolic capital. To put it another way, souvenirs are "cultural currency". Their value is derived less from their cultural and artistic significance—the cultural "gold" which supports them—and more from the demand of the market, and from their social-symbolic functions. Although souvenirs represent "cultural devaluation" and "cultural inflation" (mass production of low-face-value image currency), they nevertheless gain extra sign value in their circulation within the tourist market.

In summary, tourism is not only experiential consumption, it is also symbolic consumption which acts as a catalyst for social communication. The symbolic consumption of tourism is most typically exemplified in the consumption of souvenirs, which function as visible evidence of visitor presence at the destination. Thus, both souvenir purchase and gift giving act as a social tool of status acccumulation and "relay" the effects of tourism (i.e., the extraordinary experiences) to the post-holiday period. Under commoditization the symbolic function of souvenirs is also exploited by the tourism industry. As a result their production may cause cultural trivialization. Souvenirs thrive less on their authenticity than on their social symbolism of the exotic. In other words, souvenirs signify *seduction*, manipulated by modernity. Souvenirs constitute a sign of seduction that plays a significant role in enticing tourists to the sites of touristic consumption.

Chapter 11

Conclusion

Although there are still many more tourism issues calling for sociological study, the exploration in this book must come to an end. The chapter title "Conclusion" may be misleading, for the arguments outlined are more tentative than constituting the last word. Thus, it is better to understand "conclusion" as a temporal closure, and the purpose of this chapter is both to sum up the points of view discussed and to elaborate their implications for the sociology of tourism and the policy issues that can be derived from it.

Summary

Tourism is an indicator of the ambivalence of modernity. To understand tourism fully, it is insufficient simply to look at the external appearance of tourism. Rather, it is necessary to go beyond and beneath the phenomenon in order to discover the deeper structural and cultural forces that underlie it. Such an exploratory exercise has been one of the purposes of this book.

Tourism, of course, existed long before modernity came into being. However, it is only under modernity, particularly late modernity, that tourism has become a mass phenomenon, and hence a "total social fact" or an "international fact" (Lanfant 1995b). The conditions of modernity shape tourist motivation and demand. It is modernity that "pulls" people away from home in a quest for pleasure in other places, due to advances in technology, living standards, social welfare (e.g., travel as social right, holidays with pay), and international relations (e.g., peace and globalization). However, modernity also "pushes" people away from home to tourist destinations for relaxation, recreation, the experience of change, novelty, fantasy, and freedom. Indeed, it is because the "dark"

214

side of the structural ambivalence of modernity has become increasingly unbearable that people turn to tourism as a "way of escape" (Rojek 1993). Moreover, modernity has witnessed the emergence of the social organization and production of travel, tourism, and touristic experiences, which involve, for example, commercial transportation and communication organizations, travel agencies, tour operators, the public sector, institutional and legal issues regarding holiday and tourism (Urry 1995). In sum, modernity has transformed people into touristic subjects. The habit of travel and the motivation for tourism have become so deeply rooted in the minds of modern men and women that tourism has significantly shaped the landscape of consumer culture in modern and late modern societies. Nowadays, it is widely believed that *experiences* (e.g., travel and tourism) have become one of the largest consumer items in the West (and also in some countries in the East, such as Japan and newly industrialized countries and zones such as Singapore and Hong Kong), second only to the purchase of a house or a car (i.e., the consumption of goods). This global movement of people in the name of leisure and pleasure has already exerted, and will undoubtedly continue to exert, a whole range of economic, political, social, and cultural influences on both tourist-generating and tourist-receiving regions, so much so that most national and local governments simply cannot afford to ignore the phenomenon.

This book has attempted to theorize the phenomenon of tourism from a sociological perspective. Tourism has been considered in the context of modernity. Rather than simply accepting the commonsense notion that tourism is the result of an advance in the material conditions of modern society, this book has gone one step further and linked tourism to the wider structural and cultural contexts of modernity. Tourism can thus be considered in relation to the differentiation of Logos-modernity (where reason and rationality dominate and irrational desires and passions are constrained) and Eros-modernity (where reason and rationality recede and irrational factors are socially and culturally approved prior to their being released, consumed, and celebrated). Tourism is thus located at the confluence of Eros-modernity (*chapter 2*).

The sociogenesis of the tourist, and the social and historical conditions for the emergence of touristic motivation, are very much part and parcel of the structural ambivalence of modernity. First, the *institutional* dimension of modernity, particularly Western modernity, brings about inauthenticity. Tourism, by contrast, is a form of "authenticity-seeking" and of searching for meanings, not only in the sense of seeking authentic objects, but also in the sense of a quest for existential authenticity, by escaping from mainstream institutions and relaxing the rules of self-constraint associated with these institutions (*chapter 3*). Second, in respect of the *technological* environment of modernity, the constant advances in science and technology, which are themselves the embodiment of rationality, while protecting and supporting human life have also lowered the quality of the environment, particularly its cultural and psychological aspects. Nature tourism is a cultural and psychological reaction to this dimension of modernity (*chapter 4*). Third, the *temporal* dimension of modernity, as embodied in, for example, schedulization and routinization, is undoubtedly responsible for improved efficiency, synchonization, and order; yet it often gives rise to stress, monotony, and temporal alienation. Tourism, in the form of the institution of holidays with pay, is an alternative experience of time, i.e., time off or holiday time, which appears as an alternative rhythm, free from the constraints of the daily tempo (*chapter 5*). Fourth, the *social–spatial* dimension of modernity, as exhibited in the strategies of inclusion and exclusion, tends to exclude differences and Otherness in the form of nationalization; this is particularly evident in early Western modernity. However, nationalization also leads to the formation of national identity in the West, which paves the way to heightened cultural curiosity about differences by removing the fear of the Other. Relatedly, another major social–spatial strategy of modernity, i.e., globalization, helps people to approach the space of difference from a new perspective, such as exoticism (the love of the exotic). Ironically, however, cultural curiosity about the exotic can sometimes be a surreptitious and alternative form of neocolonialism (*chapter 6*).

There are therefore internal relationships between the aesthetic (in the sense of the *sensual*) dimension of modernity (tourism is itself an embodiment of this dimension), and the *structural* dimension of modernity (the institutional, technological, temporal, and socio-spatial dimension). The aesthetic or sensual dimension of modernity can be analyzed in terms of four fundamental categories: "authenticity", "nature", "time off", and "difference". All of these categories are exemplified in the tourism of authenticity, nature tourism, holidaying, and international tourism respectively, and each can only be understood against the structural ambivalence of modernity. Tourism, as the cultural celebration and experience of "authenticity", "nature", "time off", or "difference", is implicitly a non-verbal questioning of the existential condition of modernity. Is modernity really as liberating as the classic Enlightenment thinkers claimed? Does modernity actually lead to emancipation, without having any dehumanizing effects? The answers may not be as positive and certain as some expect. The rise and the practice of tourism is in reality an index of dissatisfaction with the dark side of structural modernity, which shapes the existential condition of modernity in a structurally ambivalent way. Modernity is in a sense "immoral". To put it another way, modernity, or institutional modernity, tends to exert "violence" on people's sensibility and experiences, that is, it rids people of the experience of authenticity, nature, differential time, and difference within the space of Logos-modernity. As a result, tourism is a claim to the "social right" of aesthetic experiences and sensibility in regard to authenticity, nature, time off, and difference. Therefore, tourism, together with leisure and other cultural practices, constitutes an aesthetic or sensual dimension of modernity, the so-called "Eros-modernity", as a necessary response and supplement to the structural dimension of modernity, or "Logos-modernity" (*chapter 2*). However, Eros-modernity (including tourism) is only possible only when structural modernity (Logos-modernity) has produced the necessary means and material conditions for people to make use of. In this sense, structural modernity is not absolutely "bad"; rather, it is structurally *ambivalent*.

Tourism emerges as a consumer activity. As mentioned above, the motivation to engage in tourism has do to with dissatisfaction with the dark side of modernity; that is, dissatisfaction with modernity gives rise to sociological motivations for tourism. However, the actual consumption of tourism involves the seductive conditions of modernity, conditions which emphasize the attractiveness of modernity. One of the strategies that modernity employs to demonstrate its superiority over traditional ways of life is that of supplying easy access to various attractions in the world, and also of turning the spaces of societies, cultures, and environments into touristic consumption sites, a trend that can be called "touristification" (Lanfant 1995b:35). By touristification is meant that each society, by seizing the "tourism opportunity space", incorporates tourism into its economic and socio-cultural systems, and—by means of commodification—transforms cultures, heritage, identities, landscapes, and townscapes that might be otherwise economically idle into commodities for touristic consumption. In brief, touristification involves both a substantial and a symbolic transformation of society into spectacles, attractions, and playgrounds. Under late modernity, the globalization of touristification indicates that modernity has transcended the spatial limits of pleasure available at home by integrating other places within the tourist reach where a wider spectrum of pleasure and enjoyment can be achieved.

Tourism is not merely a consumer activity, but also a particular consumer culture, i.e., a touristic consumer culture. Seduction by tourism is part of the general seduction taking place in consumer culture in modernity. Touristic consumer culture includes not only cultural representations of tourism, such as images and discourses, but also cultural orientations towards, and values on, the consumption of tourism, including the experiential and symbolic consumption of tourism. Consumer culture also involves the commoditization of tourism, usually carried out by the industry under the condition of globalizing capitalism. In order to attract tourists to touristic consumption sites, the industry manipulates various signs to effect the *idealization* of these destinations. Thus, images are constructed for destinations, via advertisements, for the purpose of influencing choice (*chapter 7*). Destinations are also represented as discourses, not only in advertisements but also in the wider mass media, which define tourist attractions and accordingly shape people's tastes and choices (*chapter 8*). Touristic consumer culture also involves the commoditization of travel experiences, which leads to the transformation of travel into tourism, i.e., a distinctive kind of commodity which takes the form of truncated experiences, qualitative time, or itineraries. Tourism is thus an experiential form of consumption which is culturally oriented, and to which cultural meanings are attached (*chapter 9*). It is also symbolic consumption, involving the differentiation of status, social locations, and social identities. Tourism is thus pursued as a social communication of social identity, while souvenirs are status symbols that provide visible and tangible evidence of presence in various destinations (*chapter 10*).

Tourism as Mediation in Modernity

This book has examined how an understanding of the social construction of tourism helps us to appreciate modernity in general. The sociology of tourism is, in a sense, a sociology of modernity and its associated experiences. In this book postmodernity has been treated as part of modernity, namely a dimension of late modernity. Thus, the relationship between tourism and late modernity is thought of as a specific relationship between tourism and modernity in general, while globalization is regarded as a global condition of modernity.

The implications of tourism studies go far beyond the area of tourism itself. Tourism is a mirror in which many of the "secrets" of modern life and modern existential conditions become visible and clear. Therefore, studying tourism is a particular way of studying modern life or the existential conditions of modernity. Tourism not only reveals the contradictions and ambivalence of modernity, but also holds together or mediates the two poles of a set of modern antitheses in a particular, touristic, way. Thus, the opposition between subject and object, economy and culture, base and superstructure, modernity and tradition, the global and the local, the core and the periphery are all mediated and integrated in tourism.

Tourism involves interaction and an antithesis between the touring (subject) and the toured (object). The touring, or the tourist, as an agent of modernized society, comes to be a subject. In contrast the toured, often in the form of the exotic, difference, or novelty, becomes an object of curiosity at which tourists gaze. Tourism also brings together the remote parties of the touring (subject) and the toured (object). The key medium of this encounter is money. Touring subjects offer cash to the toured (usually via brokers), and in return they consume "experiences"—the exotic, the sunny, the beautiful, and the sexy—which are organized, packaged, and sold as "itineraries". Both local suppliers and foreign brokers, such as tour operators and travel agencies in tourist-generating regions, are involved in this organization and production of tourist experiences. In addition, touring subjects who gaze at locals and their environments are themselves gazed at by the toured. Thus, both parties not only engage in commodity exchange (money in exchange for experiences), but also trade perceptions of each other. Such perceptions ("how they see us" and "how we see them") are, however, usually stereotypes. Furthermore, the gazes of both sides are not equal. It is the tourist gaze that gains ascendancy over the perception of the toured. To be gazed at is to be deprived of privacy, thus, local suppliers often present "staged authenticity" to tourists (MacCannell 1973, 1976). Both parties then play a game in which a distinctive touristic culture is produced in a touristic space.

Tourism unites economy and culture, base and superstructure in a particular way. For Marx, travel is a "secondary" consumption, based on the general surplus of social production as a whole, and on personal disposable income and time in particular. As he points out, "If I have no money for travel, I have no *need*—that is, no real and realizable need—to travel" (1977:124). In this sense, touristic activity, as a particular way of life and a temporary respite from routine roles and obligations, can be understood as a "superstructure" (Nash 1995) or a "touristic superstructure" (Nash 1981). It is a kind of leisure activity or cultural practice which depends on the economic situation of an individual or household, and mirrors cultural sanctions. Such a touristic superstructure in modernized society implies an economic opportunity for both developed and developing countries. Thus, in order to take advantage of this "opportunity space", economic bases and infrastructures must be built or improved. For local tourism suppliers, tourism is primarily, though not solely, a business. Apart from its economic importance, it is also concerned with culture, image, identity, and distinctive local features. Thus, it is a mix of economic capital and cultural capital, of utility and meaning, of economic activity and cultural activity. As Figure 11.1 shows, tourism is both an economic process and also a cultural process. It is thus an economy of culture, image, and identity. The old boundaries between economy and culture are therefore challenged by tourism.

Tourism mediates modernity and tradition on the one hand, and the local and the global on the other. From a *temporal* perspective, tourism reunites modernity and tradition. If modernization is a force of detraditionalization, that is, a force that weakens the role and position of tradition in modern society (Giddens 1994), then tourism is an agent that

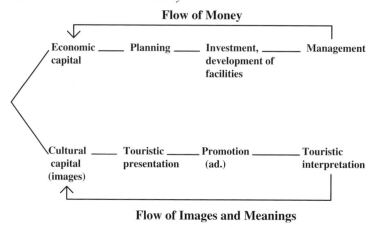

Flow of Money

Figure 11.1   **Two Interelated Processes of Tourism Production**

rediscovers the meaning and function of tradition. If modernization is characterized by homelessness, placelessness, or "de-placeness" (Cohen 1995), then tourism is a quest for "home-ness" or "placeness" via visiting "heritage", a symbol of the past, roots, and identity. Tourism, in short, is the embodiment of a nostalgia that has persisted in modernity (Dann 1994b; Turner 1987). From a *spatial* perspective, tourism integrates the local and the global. While globalization, in relation to modernity or modernization, tends to homogenize cultures, tourism appears as a search for difference, the exotic, the unusual, and the extraordinary in other places. Thus, the more locally unique and distinctive an attraction, the more globally attractive it will probably be. Tourism therefore helps incorporate the local into the global—it is a global–local nexus (Lanfant et al 1995). It thus becomes a matter of commonsense for economically weak developing countries to gain access to the "club" of international exchange by "selling" their distinctive, colorful ways of life and cultures to international tourists.

Tourism also offers a way for less developed countries of the periphery to integrate with modernized society of the core. As far as demand is concerned, tourism can be accounted for mainly in terms of modernity, especially the structural ambivalence of modernity. As far as the periphery is concerned, tourism can be considered as a development issue in the context of globalization, whereby developing countries are linked to advanced countries (Turner and Ash 1975). If the demand for tourism in the core is a *cultural* response to the existential condition of modernity, then tourism development in the periphery is an *economic and political* response to the conditions of globalization, or globalizing modernity, of which international tourism is a part. Tourism therefore brings the agents of the core and the periphery into closer communication. Tourism is a new form of "North–South dialogue", or a "meta-tourism", structured by unequal global conditions (Burns and Holden 1995). It is an encounter between the agents of modernized society and the agents of developing society.

As is well documented, one of the difficulties for the developing countries in this "dialogue" is that tourism is likely to result in dependency on and vulnerability to fluctuations in the core. Another difficulty is that tourist demand and tourism supply often follow a different logic. The logic of tourist demand is consumerism, hedonism, and materialism, which, when extended to developing countries that are still economically weak and culturally vulnerable, may have negative effects on local communities and cultures. If for some developing countries tourism signifies opportunity, then for others it creates a "Pandora's Box" which, once opened, releases "evils" unknown, such as increasing crime, a weakening sense of community, and environmental degradation. Indeed for some developing countries, to develop or not to develop tourism is still a serious dilemma. To deal with such problems, an increasingly normative approach has developed in scholarly circles, as exhibited in the widespread popularity of the expressions "alternative tourism" and "sustainable tourism" in academic discourse (Murphy 1994; Nelson, Butler and Wall 1993). Such a normative requirement for tourism development must, however, also be the concern of both local agents (including planners and policy makers) and foreign agents (such as tourists, investors, and international organisations—Lea 1993). It should be *reflexive*, in the sense that it must be well planned, based on scientific and democratic decision making, local community participation, and timely feedback. It should be *sustainable*, in the sense that it must not threaten, damage, or destroy the very resources upon which it relies. It should be *just and moral*, in the sense that tourists, policy makers, planners, and developers should show respect for the well-being of local residents, local cultures, and the local environment (Hultsman 1995).

From Problematic Tourism to Humanistic Tourism

Compared with the management literature on tourism, the sociology of tourism seems to be less practical. However, this is not to say that the sociology of tourism is insignificant. Rather, it reveals the societal and cultural aspects of tourism in terms of the larger context of society, an essential ingredient in identifying the roots of the current problems with tourist practices. The sociology of tourism is useful not only because it offers a deeper and wider understanding of tourism, but also because it illuminates the fundamental principles that should underlie the development of tourism and tourism policy. Thus, in this section the sociology of tourism is brought to bear on current policy problems and the principles and philosophy which should underlie them.

Tourism not only involves the structural ambivalence of modernity, but also faces dilemmas of its own (Plog 1994). Tourism is an escape from the alienation of modernity. However, it may sometimes fall prey to its own alienation—touristic alienation. For example the industry may encourage the overschedulization of itineraries, crowding, long queues in traffic, and cheating by locals. The consequences of touristic consumption may also be problematic. While tourism is a quest to discover nature and exotic cultures, it may at the same time place them at risk. Whereas tourism is the pursuit of an elementary way of life, the "search for simplicity ends in technological complexity and accelerated social change; the pursuit of the exotic and diverse ends in uniformity" (Turner and Ash 1975:292). These dilemmas are demonstrated most seriously in mass tourism. That is why the voice of negative criticism has been frequently heard. Some conservationists even suggest that tourism should be banned indefinitely.

However, a total ban on tourism would be unwise and unviable. Tourists should have a right to travel, and to see landscapes and cultures in other places. To ban tourism would also lead to a waste of resources. What is at stake here is not whether tourism should be developed, but rather *how* it should be developed, and how its problematic consequences can be prevented. In response to such questions, "alternative tourism" or "sustainable tourism" have become very popular in the 1990s. If tourism is the result of people's cultural reaction to the existential conditions of modernity and globalization, then, "alternative tourism" is the result of people's (both tourists' and suppliers') critical response to the negative consequences and impact of mass tourism. That is why the contemporary tourism trade has recently undergone several changes (Brohman 1996; Burns and Holden 1995; Butler and Pearce 1995; Nelson et al 1993; Smith and Eadington 1992; Urry 1990a, 1994a).

Some of these modifications can be easily identified, such as the transition from Fordist to post-Fordist patterns of tourism *(chapter 5)*. Other alterations are less visible. Even so, they represent the future trend in tourism since they are based on new and sound principles and philosophy, derived from the lessons and problems of the past. These principles and philosophy can be best elucidated in terms of the sociology of tourism and they have policy implications for tourism development in the future. According to Hall, "tourism public policy is whatever governments choose to do or not to do with respect to tourism" (1995:7-8). Governments will have to identify future trends and the associated principles and philosophy if they are to make sound policies. The changes that represent the future trend in tourism on the basis of new principles and philosophy are identified as follows.

From Short-Termist Tourism to Sustainable Tourism

Mass tourism, as a Fordist pattern of tourism, is criticized as problematic. This is understandable. However, at a deeper level, what underlies the fact that mass tourism often gives rise to negative consequences is not the pattern itself (i.e., Fordism), but the realization that tourism production is often informed by *instrumentalism* and *short-termism,* which disregard the interests and long-term well-being of the host community, the local culture, and the environment. Therefore, in order to minimize these effects, tourism development should entail not merely a change of *pattern,* but more importantly a change of *philosophy,* i.e., a shift from short-termism to long-termism, from instrumentalism to humanism, and from one-sided to all-encompassing policies. Thus, the long-term interests of destinations, including those of the community and the environment, rather than prospective short-term economic benefits, should be given priority (Murphy 1985). Furthermore, policy makers and planners should display a humanistic concern not only for tourists but also for hosts, and such a concern should gain ascendancy over the instrumental purpose of tourism, i.e., the profits earned by business organizations. Finally, tourism should be integrated within the host community, a situation in which both parties are winners (Hall 1994).

This change of philosophy has already occurred in some academic circles and business groups (Burns and Holden 1995:ch.8; Nelson et al 1993). However, it should also be reflected in public policy-making and planning with respect to tourism. In other words, "sustainable tourism" should become the new direction for development, not only in advanced countries, but also in developing countries during this decade and beyond (Brohman 1996; Nelson et al 1993).

However, different regions have quite different policy concerns for the sustainability of tourism. In advanced countries, a well-developed economy and social structure allow them to show a high policy concern for environmental quality and the sustainability of tourism. As a result, tourism is comparatively better planned, better managed, and better integrated into the community as a whole. By contrast, in some developing countries critical economic situations (e.g., poverty) tend to make policy makers and planners prioritize economic considerations at the expense of sustainability.

Sustainable tourism should be understood not only in environmental terms, but also, more importantly, in economic and structural terms. In many developing countries, to overcome the shortage of development funds foreign capital is introduced. However, in a context where public policy concern for the sustainability of tourism is lacking, foreign investors and tour operators tend to adopt a short-term approach, at the expense of the long-term well-being of the destination community and sustainable development, not only environmentally (e.g., environmental damage) but also economically (e.g., foreign exchange leakage).

In sum, sustainable tourism is particularly significant for developing countries, because it is in these countries that the negative consequences of mass tourism, partly as a result of "wrong" policies and partly because of the asymmetrical nature of the global political economy, is most often found, such as environmental damage, foreign exchange leakage, decline of community integration, and crises of identity. While acknowledging that tourism development strategies should vary with different conditions, Brohman (1996) has suggested five alternative strategies for Third World or developing countries. First, alternative tourism should consist of smaller-scale, dispersed, and low-density developments (post mass tourism). Second, there should be more local ownership, often family-owned, relatively small-scale businesses, rather than foreign-owned large-scale enterprises (as in mass tourism). This would help increase the multiplier effect and spread benefits within the host community; it would also help prevent excessive foreign exchange leakage. Third, there should be more community participation in tourism planning at the local or regional level. Fourth, there should be greater stress on sustainability in both the environmental and the cultural sense. Finally, alternative tourism should be aimed at preventing any destruction of the host culture.

From Irresponsible Tourism to Responsible Tourism

As previously noted, tourism was initially an attempt to escape from the dark side of modernity. The latter is related to the underlying principle of modernity—"the primacy of instrumental reason" (Taylor 1991). The problematic of mass tourism also has to do with the extension of this underlying principle to the area of *tourism production*, since business organizations who supply the products of tourism are first and foremost informed by the "primacy of instrumental reason", rather than by humanistic concerns. Therefore, to paraphrase Fromm, they are preoccupied more with the question "what is good for the business and profit?" than with the question "what is good for both tourists and the host community?" (1976:16).

Overcoming the disadvantageous consequences of irresponsible tourism requires a shift in philosophy in the running of tourist businesses, and greater involvement by consumer groups, local people, and the state.

First, for entrepreneurs and brokers, the philosophy that underlies supply should not be the "primacy of instrumental reason"—a principle that is related to the "modernity problematic"—but rather an ethic of responsibility, whereby both tourists and local hosts are no longer treated as the mere *means* to profit, but rather as humans whose long-term well-being should be respected and protected. This does not imply that businessmen should no longer make any money; rather, that the profits derived from tourism should not be obtained at the cost of tourists, local people, and the environment. Second, the market itself does not automatically ensure a humanistic concern or a sense of responsibility for hosts because it is the consumers who pay the entrepreneurs and brokers, and receive their quality service and humanistic concern. Thus, all too often a local host's well-being is sacrificed in favor of the satisfaction of consumers. In response to this situation, various supervizing agents and mechanisms should be put in place to protect the interests of local hosts. The state is one such agent. Another is the new social movement of "Tourism Concern", a new community interest group that is concerned with the growth and impact of tourism (Botterill 1991). Tourism development is thus not merely an economic issue; it is often a political issue (Hall 1994; Richter 1989). Tourism is, in a sense, a game revolving around "who wins, and who loses" (Hall 1994). To ensure that it becomes a game in which everybody wins, an ethic of responsibility and humanistic concern must be adopted by governments and embodied in public policy. Tourism should be conceived of in humanistic rather than purely economic terms. The development of humanistic and responsible tourism will help tourists to satisfy their desires in response to the "modernity problematic", and the host community to benefit from tourism under the condition of globalization.

Responsible tourism should be the concern of not only planners and policy makers, but also tourists themselves (Lea 1993). For local planners and policy makers, some negative aspects of tourism are, to a certain extent, unavoidable, such as "cultural superficialization". There are also some factors that are beyond the control of the planners, for example the exchange rate, the misconduct of tourists, and the spread of AIDS via sex tourism. Therefore, responsible tourism entails the cooperation of both tourists and tourist brokers. For example, certain forms of tourism tend to have a greater negative impact than others, such as those which involve child prostitution and drug smuggling, activities which should be banned or restricted. Tourists should not be encouraged by operators to engage in these kinds of tourism. Tourism, as a lifestyle, *should be* sustainable (Leslie 1994). At the root of this sustainability is the necessity for the tourist lifestyle to be moral. In some cases tourism involves as an "anticultural" (Turner and Ash 1975) ritual (e.g., nudity on the beach). If this behavior is prohibited in sober Western societies, it is clearly immoral and offensive to act in such a manner when visiting traditional societies which, for the sake of economic benefit, have to "tolerate" such a conduct.

How tourism will evolve in the future is uncertain. Some analysts have argued that the post-tourist will replace the tourist and that post-tourism will replace tourism, because the boundaries between tourist and non-tourist, or between tourism and non-tourism, have been or will be blurred (Fiefer 1985; Urry 1990a). Whether or not this situation is, or will be, a reality is still a matter of debate. However, the conclusion of this book is that future tourism needs to be based on fundamental and internationally agreed general principles and philosophy which display humanistic concern for both the tourist and the host society. These principles and philosophy need to be embodied in future policies and regulatory frameworks for national and international tourism.

References

Adam, Barbara
1990   Time and Social Theory. Cambridge: Polity Press.
Adams, Kathleen M.
1984   Come to Tana Toraja, "Land of the Heavenly Kings": Travel Agents as Brokers in Ethnicity. Annals of Tourism Research 11:469–85.
1989a   Origins of Sightseeing, Annals of Tourism Research 16(1):7–29.
Adler, Judith
1989b   Travel as Performed Art. American Journal of Sociology 94 (6):1366–1391.
Albers, Patricia C., and William R. James
1988   Travel Photography: A Methodological Approach. Annals of Tourism Research 15:134:158.
Allcock, John B.
1988   Tourism as Sacred Journey. Loisir et Société, 11(1):33–48.
1989   Sociology of Tourism. In Tourism Marketing and Management Handbook, Stephen F. Witt and Luiz Moutinho, eds., pp. 407–414. London: Prentice Hall.
1995   International Tourism and the Appropriation of History in the Balkans. In International Tourism: Identity and Change, M.-F. Lanfant, J. B. Allcock and E. M. Bruner, eds., pp. 100–112. London: Sage.
Alter, Peter
1985   Nationalism. London: Edward Arnold.
Anderson, Benedict
1983   Imagined Communities: Reflections on the Origin and Spread of Nationalism (revised ed.). London: Verso.
Anderson, Luella F. and Mary Ann Littrell
1995   Souvenir-Purchase Behavior of Women Tourists. Annals of Tourism Research 22(2):328–348.
Andrews, Malcolm
1989   The Search for the Picturesque: Landscape Aesthetics and Tourism in Britain, 1760–1800. Aldershot: Scolar Press.
Apostolopolous, Yiorgos, Stella Leivadi and Andrew Yiannakis, eds.
1996   The Sociology of Tourism: Theoretical and Empirical Investigations. London: Routledge.
Appadurai, Arjun
1990   Disjunction and Difference in the Global Cultural Economy. Theory, Culture & Society 7:295–310.
Apter, Michael J.
1992   The Dangerous Edge: the Psychology of Excitement. New York: Free Press.

Aristotle
  1925   The Nicomachean Ethics. Trans. by David Ross and revised by J. L.
         Ackrell and J. O. Urmson. Oxford: Oxford University Press.
Arnason, Johann P.
  1990   Nationalism, Globalisation and Modernity. Theory, Culture & Society
         7:207–236.
Ashworth, G. J. and H. Voogd
  1990   Selling the City. London: Belhaven Press.
Barke, Michael and Ken Harrop
  1994   Selling the Industrial Town: Identity, Image and Illusion. *In* Place
         Promotion: the Use of Publicity and Marketing to Sell towns and Regions,
         John R. Gold and Stephen V. Ward, eds., pp. 93–114. Chichester: Wiley.
Barthes, Roland
  1972:   Mythologies. London: Jonathan Cape.
Baudrillard, Jean
  1983   Simulations. Trans. by Paul Foss, Paul Patton and Philip Beitchman.
         New York: Semiotext(e).
  1988   Selected Writings. Ed. by M. Poster. Cambridge: Polity Press.
Bauman, Zygmunt
  1987   Legislators and Interpreters: on Modernity, Post-modernity and
         Intellectuals. Cambridge: Polity Press.
  1989   Modernity and the Holocaust. Cambridge: Polity Press.
  1990a   Modernity and Ambiguity. Theory, Culture & Society 7:143–169.
  1990b   Thinking Sociologically. Oxford: Blackwell.
Beck, Ulrich
  1992   Risk Society: towards a New Modernity. Trans. by Mark Ritter. London:
         Sage.
Berger, John
  1972   Ways of Seeing. London: British Broadcasting Corporation and
         Penguin Books.
Berger, Peter L.
  1973   "Sincerity" and "Authenticity" in Modern Society. Public Interest 31
         (Spring):81–90.
Berger, Peter L., Brigitte Berger and Hansfried Kellner
  1973   The   Homeless   Mind: Modernization   and   Consciousness.
         Harmondsworth: Penguin.
Berger, P. L. and T. Luckmann
  1971   The Social Construction of Reality: A Treatise in the Sociology of
         Knowledge. Harmondsworth: Penguin.
Berlyne, D. E.
  1960   Conflict, Arousal and Curiosity. New York: McGraw-Hill.
  1966a   Conflict and Arousal. Scientific American 215:82–87.
  1966b   Curiosity and Exploration. Science 155:25–33.
Berman, M.
  1970   The Politics of Authenticity. London: Allen & Unwin.
Bernstein, J. M.
  1991   Introduction. *In* T. W. Adorno: The Cultural Industry, edited by J. M.
         Bernstein, pp.1–25. London: Routledge.
Beynon, H.
  1973   Working for Ford. Harmondsworth: Penguin.
Blauner, Robert
  1964   Alienation and Freedom. Chicago: University of Chicago Press.

Boissevain, Jeremy (ed.)
1996 Coping with Tourists: European Reactions to Mass Tourism. Oxford: Berghahn.
Bojanic, David C.
1991 The Use of Advertising in Managing Destination Image. Tourism Management 12(4):352–55.
Boniface, Priscilla and Peter J. Fowler
1993 Heritage and Tourism in "the Global Village". London: Routledge.
Boorstin, D.
1964 The Image: a Guide to Pseudo-events in America. New York: Harper & Row.
Böröcz, József
1996 Leisure Migration: A Sociological Study on Tourism. Oxford: Elsevier Science.
Botterill, T. David
1991 A New Social Movement: Tourism Concern, the First Two Years. Leisure Studies, 10: 203–217.
Bourdieu, Pierre
1977 Outline of a Theory of Practice. Cambridge: Cambridge University Press.
1984 Distinction: a Social Critique of the Judgement of Taste. Trans. By Richard Nice. London: Routledge.
1990 Time Perspectives of the Kabyle. *In* The Sociology of Time. J. Hassard, ed., p. 219–237. London: Macmillan.
Bowie, Andrew
1990 Aesthetics and Subjectivity: from Kant to Nietzsche. Manchester: Manchester University Press.
Bramwell, Bill and Liz Rawding
1996 Tourism Marketing Image of Industrial Cities. Annals of Tourism Research 23(1):201–221.
Braverman, H.
1974 Labor and Monopoly Capital. New York: Monthly Review Press.
Brendon, P.
1991 Thomas Cook: 150 Years of Popular Tourism. London: Secker & Warburg.
Britton, R. A.
1979 The Image of the Third World in Tourism Marketing. Annals of Tourism Research 6(3):318–29.
Britton, S.
1982 The Political Economy of Tourism in the Third World. Annals of Tourism Research 9:331–359.
1991 Tourism, Capital, and Place: towards a Critical Geography of Tourism. Environment and Planning D: Society and Space 9:451–478.
Brohman, John
1996 New Directions in Tourism for Third World Development. Annals of Tourism Research 23 (1):48–70.
Brown, David
1996 Genuine Fakes. *In* The Tourist Image: Myths and Myth Making in Tourism, Tom Selwyn, ed., pp. 33–47. Chichester: Wiley.
Brown, Graham
1992 Tourism and Symbolic Consumption. *In* Choice and Demand in Tourism, Peter Johnson and Barry Thomas, eds., pp. 57–71. London: Mansell.

Brown, Hilton
    1950   To Get Away from It All. *In* Traveler's Quest, A. Michael, ed., pp.272–290. London: William Hodge.
Bruner, Edward M.
    1989   Tourism, Creativity, and Authenticity. Studies in Symbolic Interaction 10:109–114.
    1991   Transformation of Self in Tourism. Annals of Tourism Research 18(2): 238–250.
    1994   Abraham Lincoln as Authentic Reproduction: A Critique of Postmodernism. American Anthropologist 96(2):397–415.
    1995   The Ethnographer/Tourist in Indonesia. *In* International Tourism: Identity and Change, Marie-Francoise Lanfant, John B. Allcock and Edward M. Bruner, eds., pp. 224–241. London: Sage.
Buck, R.
    1977   The Ubiquitous Tourist Brochure. Explorations in Its Intended and Unintended Use. Annals of Tourism Research 4:195–207.
Burawoy, Michael
    1979   Manufacturing Consent: Changes in the Labor Process under Capitalism. Chicago: University of Chicago Press.
Burns, Peter M. and Andrew Holden
    1995   Tourism: a New Perspective. London: Prentice Hall.
Burton, Suzy Kruhse-Mount
    1995   Sex Tourism and Traditional Australian Male Identity. *In* International Tourism: Identity and Change, M.-F. Lanfant, J. B. Allcock and E. M. Bruner, eds., pp.192–204. London: Sage.
Butler, Richard, and Thomas Hinch, eds.
    1996   Tourism and Indigenous People. London: International Thomson Business Press.
Butler, Richard and Douglas Pearce, eds.
    1995   Change in Tourism: People, Places, Processes. London: Routledge
Buttimer, Anne and David Seamon, eds.
    1980   The Human Experience of Space and Place. London: Croom Helm.
Buzard, James
    1993   The Beaten Track: European Tourism, Literature, and the Ways to Culture, 1800–1918. Oxford: Clarendon Press.
Campbell, Colin
    1987   The Romantic Ethic and the Spirit of Modern Consumerism. Oxford: Blackwell.
    1995   The Sociology of Consumption. *In* Acknowledging Consumption: A Review of New Studies, Daniel Miller, ed., pp. 96–126. London: Routledge.
Camus, Albert
    1955   The Myth of Sisyphus. Trans. by Justin O'Brien. London: Hamish Hamilton.
Castells, Manuel
    1977   The Urban Question. London: Edward Arnold.
    1996   The Information Age: Economy, Society and Culture, Vol. I: The Rise of the Network Society. Oxford: Blackwell.
Chalfen, Richard M.
    1979   Photography's Role in Tourism: Some Unexplored Relationships. Annals of Tourism Research 6(4):435–447.
Chinoy, E.
    1955   Automobile Workers and American Dream. New York: Beacon Press.

Chon, Kye-Sung
  1990 The Role of Destination Image in Tourism: A Review and Discussion. The Tourist Review 45(2):2–9.
Cleverdon, R.
  1979 The Economic and Social Impact of International Tourism on Developing Countries. London: Economic Intelligence Unit.
Cohen, Erik
  1972 Towards a Sociology of International Tourism. Social Research 39(1):164–182.
  1974 Who is a Tourist?: a Conceptual Clarification. Sociological Review 22:527–555.
  1979a Rethinking the Sociology of Tourism. Annals of Tourism Research 6(1):18–35.
  1979b A Phenomenology of Tourist Experiences. Sociology 13(2): 179–201.
  1984 The Sociology of Tourism: Approaches, Issues, and Findings. Annual Review of Sociology 10:373–392.
  1985 Tourism as Play. Religion 15:291–304.
  1988a Traditions in the Qualitative Sociology of Tourism. Annals of Tourism Research 15(1):29–46.
  1988b Authenticity and Commoditization in Tourism. Annals of Tourism Research 15(3):371–386.
  1993 Introduction: Investigating Tourist Arts. Annals of Tourism Research 20(1):1–8.
  1995 Contemporary Tourism – Trends and Challenges: Sustainable Authenticity or Contrived Post-modernity? *In* Change in Tourism: People, Places, Progresses, Richard Butler and Douglas Pearce, eds., pp.12–29. London: Routledge.
Cohen, Stanley and Laurie Taylor
  1992 Escape Attempts: the Theory and Practice of Resistance to Everyday Life (2nd ed.) London: Routledge.
Cook, Guy
  1992 The Discourse of Advertising. London: Routledge.
Cooper, Cary L. and Judi Marshall, eds.
  1980 White Collar and Professional Stress. Chichester: Wiley.
Cooper, Cary L. and Roy Payne, eds.
  1978 Stress at Work. Chichester: Wiley.
Cooper, Cary L. and Roy Payne
  1980 Current Concerns in Occupational Stress. Chichester: Wiley.
Cooper, Chris, John Fletcher, David Gilbert, and Stephen Wanhill
  1993 Tourism: Principles & Practice. London: Pitman.
Corbin, Alain
  1994 The Lure of the Sea: the Discovery of the Seaside in the Western World 1750–1840. Trans. by Jocelyn Phelps. Harmondsworth: Penguin.
Corrigan, Peter
  1997 The Sociology of Consumption: An Introduction. London: Sage.
Coser, Lewis A. and Rose L. Coser
  1990 Time Perspective and Social Structure. *In* The Sociology of Time, J. Hassard, ed., p. 191–202. London: Macmillan.
Costa, Janeen Arnold
  1993 Tourism as Consumption Precipitate: An Exploration and Example. *In* European Advances in Consumer Research, vol.1, W. Fred van Raaij and

Gary J. Bamossy, eds., pp. 300–306. Provo, UT: Association for Consumer Research.

Coupland, Nikolas and Justine Coupland
1997 Bodies, Beaches and Burn-Times: "Environmentalism" and Its Discursive Competitors. Discourse & Society 8(1):7–25.

Crawshaw, Carol and John Urry
1997 Tourism and the Photographic Eye. *In* Touring Cultures: Transformations of Travel and Theory, Chris Rojek and John Urry, eds., pp. 176–195. London: Routledge.

Crick, Malcolm
1989 Representations of Sun, Sex, Sights, Savings and Servility: International Tourism in the Social Sciences. Annual Review of Anthropology 18:307–44.
1995 The Anthropologist as Tourist: an Identity in Question. *In* International Tourism: Identity and Change, M.-F. Lanfant, J. B. Allcock and E. M. Bruner, eds., pp. 205–23. London: Sage.

Crompton, John L.
1978 An Assessment of the Image of Mexico as a Vacation Destination and the Influence of Geographical Location upon that Image. Journal of Travel Research 18(3):18–23.

Csikszentmihalyi, Mihaly
1975 Beyond Boredom and Anxiety: the Experience of Play in Work and Games. Oxford and San Francisco: Jossey-Bass.
1988 The Flow Experience and Its Significance in Human Psychology. *In* Optimal Experience: Psychological Studies of Flow in Consciousness, M. Csikszentmihalyi and I. S. Csikszentmihalyi, eds. Cambridge: Cambridge University Press.

Csikszentmihalyi, Mihaly and Isabella Selega Csikszentmihalyi, eds.
1988 Optimal Experience: Psychological Studies of Flow in Consciousness. Cambridge: Cambridge University Press.

Csikzentmihalyi, Mihaly and Eugene Rochberg-Halton
1981 The Meaning of Things: Domestic Symbols and the Self. Cambridge: Cambridge University Press.

Culler, Jonathan
1981 Semiotics of Tourism. American Journal of Semiotics 1(1–2):127–140.

Daniel, Yvonne Payne
1996 Tourism Dance Performances: Authenticity and Creativity. Annals of Tourism Research, 23(4):780–797.

Dann, Graham M. S.
1976 The Holiday was Simply Fantasic. Revue de Tourisme 3:19–23.
1977 Anomie, Ego-enhancement and Tourism. Annals of Tourism Research 4:184–94.
1981 Tourist Motivation: an Appraisal. Annals of Tourism Research 8:187–219.
1988 Images of Cyprus Projected by Tour Operators. Problems of Tourism XI (3):43–70.
1989 The Tourist as Child: Some Reflections. Cahiers du tourisme, Serie C. No. 135. Aix-en-Provence: CHET.
1993 Advertising in Tourism and Travel: Tourism Brochures. *In* VNR's Encyclopedia of Hospitality and Tourism, M. Khan, M. Olsen and T. Var, eds., pp. 893–901. New York: Van Nostrand Reinhold.
1994a A Sociolinguistic Analysis of the Cognitive, Affective and Conative Content of Images as Alternative Means to Gauging Tourist Satisfaction:

Motivation and Experience. *In* Spoilt for Choice: Decision Making Processes and Preference Changes of Tourists—Intertemporal and Intercountry Perspectives, R. Gasser and K. Weiermair, eds., pp.125–139. Thaur: Kulturverlag.

1994b   Tourism: the Nostalgia Industry of the Future. *In* Global Tourism: the Next Decade, W. Theobald, ed., pp. 55–67. London: Butterworth-Heinemann.

1996a   The Language of Tourism: A Sociolinguistic Perspective. Wallingford: CAB International.

1996b   The People of Tourist Brochures. *In* The Tourist Image: Myths and Myth Making in Tourism, Tom Selwyn, ed., pp. 61–81. Chichester: Wiley.

1996c   Images of Destination People in Travelogues. *In* Tourism and Indigenous People, R. Butler and T. Hinch, eds., pp. 349–375. London: International Thomson Business Press.

1997   The Green Green Grass of Home: Nature and Nurture in Rural England. *In* Tourism, Development and Growth: the Challenge of Sustainability. Salah Wahab and John J. Pigram, eds., pp. 257–273. London: Routledge.

Dann, Graham M. S. and Erik Cohen
    1991   Sociology and Tourism. Annals of Tourism Research 18(1):155–169.

de Kadt, E.
    1979   Tourism: Passport to Development? Oxford: Oxford University Press.

Dilley, Robert S.
    1986   Tourist Brochures and Tourist Images. Canadian Geographer 30(1):59–65.

Dimanche and Samdahl
    1994   Leisure as Symbolic Consumption: a Conceptualization and Prospectus for Future Research. Leisure Studies 16(2):119–129.

Dondis, Donis
    1973   A Primer of Visual Literacy. Cambridge: MIT Press.

Douglas, Mary and Baron Isherwood
    1996   The World of Goods: towards an Anthropology of Consumption. London: Routledge.

Dovey, K.
    1985   The Quest for Authenticity and the Replication of Environmental Meaning. *In* Dwelling, Place and Environment, D. Seamon and R. Mugerauer, eds., pp. 33–49. Lancaster: Martinus Nijhoff.

Dumazedier, Joffre
    1967   Toward a Society of Leisure. New York: Free Press.

Duncan, J. S.
    1978   The Social Construction of Unreality: an Interactionist Approach to the tourist's Cognition of Environment. *In* Humanistic Geography, David Ley and Marwyn S. Samuels, eds., pp. 269–282. Chicago: Maaroufa Press.

Durkheim, Emile
    1964   The Division of Labor in Society. New York: Free Press.
    1995   The Elementary Forms of Religious Life. Trans. by Karen E. Fields. New York: Free Press.

Dyer, Gillian
    1982   Advertising as Communication. London: Methuen.

Eco, U.
    1986   Travels in Hyperreality. Trans. by William Weaver. London: Picador.

Edwards, Elizabeth
 1996 Postcards—Greetings from Another World. *In* The Tourist Image: Myths and Myth Making in Tourism, Tom Selwyn, ed., pp. 197–221. Chichester: Wiley.
Ehrentraut, A.
 1993 Heritage Authenticity and Domestic Tourism in Japan. Annals of Tourism Research 20(2):262–278.
Elias, Norbert
 1956 Problems of Involvement and Detachment. British Journal of Sociology 7(3):226–252.
 1978 The Civilizing Process, Vol. 1: the History of Manners. Oxford: Blackwell.
 1982 The Civilizing Process, Vol. 2: State Formation and Civilization. Oxford: Blackwell.
 1992 Time: an Essay. Oxford: Blackwell.
Elias, Norbert and Eric Dunning
 1986 Quest for Excitement: Sport and Leisure in the Civilizing Process. Oxford: Blackwell.
Entrikin, J. Nicholas
 1991 The Betweenness of Place: toward a Geography of Modernity. London: Macmillan.
Fairclough, Norman
 1989 Language and Power. London: Longman.
Featherstone, Mike, ed.
 1990 Global Culture. London: Sage.
 1991a The Body in Consumer Culture. *In* The Body: Social Process and Cultural Theory. M. Featherstone, M. Hepworth and B. S. Turner, eds., pp. 170–196. London: Sage.
 1991b Consumer Culture and Postmodernism. London: Sage.
 1995 Undoing Culture: Globalisation, Postmodernism and Identity. London: Sage.
Featherstone, M. M. Hepworth and B. S. Turner, eds.
 1991 The Body: Social Process and Cultural Theory. London: Sage.
Featherstone, Mike, Scott Lash and Roland Robertson, eds.
 1995 Global Modernities. London: Sage.
Fenickel, O.
 1951 On the Psychology of Boredom. *In* Organization of Pathology and Thought, D. Rapaport, ed. New York: Columbia University Press.
Fiefer, M.
 1985 Going Places. London: Macmillan.
Fjellman, S. M.
 1992 Vinyl Leaves: Walt Disney World and America. Boulder: Westview Press.
Forgays, Donald G., Tytus Sonowiski and Kazimierz Wrzesniewski, eds.
 1992 Anxiety: Recent Developments in Cognitive, Psychophysiological, and Health Research. Washington: Hemisphere.
Fornäs, Johan
 1995 Cultural Theory & Late Modernity. London: Sage.
Foster, J.
 1964 The Sociological Consequences of Tourism. International Journal of Comparative Sociology 5:217–227.

Foucault, Michel
   1976   The Birth of the Clinic. London: Tavistock.
   1977   Discipline and Punish: the Birth of the Prison. Harmondsworth: Penguin.
   1989   The Archaeology of Knowledge. Trans. by A. M. Sheridan Smith. London: Routledge.
Fox, Roy F.
   1994a Introduction. *In* Images in Language, Media, and Mind. R. F. Fox, ed., pp. ix–xiii. Urbana, Illinois: National Council of Teachers of English.
Fox, Roy F., ed.
   1994b, Images in Language, Media, and Mind. Urbana, Illinois: National Council of Teachers of English.
Freud, S.
   1963   Civilization and its Discontents. Trans. by J. Riviere, revised and newly edited by J. Strachey. London: The Hogarth Press and the Institute of Psycho-analysis.
Friedland, Roger and Deirdre Boden
   1994   NowHere: an Introduction to Space, Time and Modernity. *In* NowHere: Space, Time and Modernity, R. Fredland and D. Boden, eds., pp. 1–60. London: University of California Press.
Fromm, Erich
   1956   The Sane Society. London: Routledge & Kegan Paul.
   1976   To Have or to Be? London: ABACUS.
Frow, John
   1991   Tourism and the Semiotics of Nostalgia. October 57:123–151.
Furbank, P. N.
   1970   Reflections on the Word "Image". London: Secker & Warburg.
Furst, Lilian R.
   1976   Romanticism (2nd ed.) London: Methuen.
Fussell, P.
   1980   Abroad: British Literary Traveling between the Wars. Oxford: Oxford University Press.
Fyfe, Gorden and John Law
   1988   Introduction: On the Invisibility of the Visual. *In* Picturing Power: Visual Depiction and Social Relations, G. Fyfe and J. Law, eds., pp. 1–14. London: Routledge.
Gadamer, Hans-Georg
   1976   Philosophical Hermeneutics. Trans. and ed. by David E. Linge. Berkeley: University of California Press.
Galbraith, John Kenneth
   1958   The Affluent Society. London: Hamish Hamilton.
Garson, Barbara
   1975   All the Livelong Day: the Meaning and Demeaning of Routine Work. New York: Doubleday.
Gartner, William C.
   1993   Image Formation Process. Journal of Travel and Tourism Marketing 2(2/3):191–215.
Gartner, William C. And John D. Hunt
   1987   An Analysis of State Image Change over a Twelve-Year Period (1971–1983). Journal of Travel Research 26(2):15–19.

Geiwitz, P. J.
  1966   Structure of Boredom. Journal of Personality and Social Psychology 3:592–600.
Gergen, K. J.
  1985   The Social Constructionist Movement in Modern Psychology. American Psychologist 40:266–275.
Gergen, K. J. and M. M. Gergen
  1991   Toward Reflexive Methodologies. *In* Research and Reflexivity. F. Steier, ed., pp. 76–95. Newbury Park, CA: Sage.
Giddens, Anthony
  1976   New Rules of Sociological Method: A Positive Critique of Interpretative Sociologies. London: Hutchinson.
  1979   Central Problems in Social Theory: Action, Structure and Contradiction in Social Analysis. London: Macmillan.
  1981   A Contemporary Critique of Historical Materialism. London: Macmillan.
  1984   The Constitution of Society: Outline of the Theory of Structuration. Cambridge: Polity Press.
  1989   Sociology. Cambridge: Polity Press.
  1990   The Consequences of Modernity. Cambridge: Polity Press.
  1991   Modernity and Self-identity: Self and Society in the Late Modern Age. Cambridge: Polity Press.
  1994   Modernity and De-traditionalisation. *In* Reflexive Modernisation, U. Beck, A. Giddens and S. Lash, eds., pp. 56–109. Cambridge: Polity Press.
Gilbert, D. C.
  1990   Conceptual Issues in the Meaning of Tourism. *In* Progress in Tourism, Recreation and Hospitality Management, vol. 2, C. P. Cooper, ed., pp. 4–27. London: Behaven Press.
  1994   The European Community and Leisure Lifestyles. *In* Progress in Tourism, Recreation and Hospitality Management, vol. 5, C. P. Cooper and A. Lockwood, eds., pp. 116–131. Chichester: Wiley.
Goffman, Erving
  1959   The Presentation of Self in Everyday Life. Harmondsworth: Penguin.
Gold, John R.
  1994   Locating the Message: Place Promotion as Image Communication. *In* Place Promotion: the Use of Publicity and Marketing to Sell Towns and Regions, John R. Gold and Stephen V. Ward, eds., pp.19–37. Chichester: Wiley.
Goldberger, Leo and Shlomo Breznitz, eds.
  1993   Handbook of Stress: Theoretical and Clinical Aspects (2nd ed.) New York: Free Press.
Goldman, Robert
  1992   Reading Ads Socially. London: Routledge.
Goldthorpe, J. H., D. Lockwood and Colleagues
  1968   The Affluent Worker: Industrial Attitudes and Behaviour. Cambridge: Cambridge University Press.
Golomb, Jacob
  1995   In Search of Authenticity. London: Routledge.
Gordon, Beverly
  1986   The Souvenir: Messenger of the Extraordinary. Journal of Popular Culture 20(3):135–146.

Gottlieb, A.
1982   American's Vacations. Annals of Tourism Research 9:165–187.
Graburn, Nelson H. H., ed.
1976   Ethnic and Tourist Arts. Cultural Expressions from the Fourth World. Berkeley: University of California Press.
Graburn, Nelson H. H.
1983a   The Anthropology of Tourism. Annals of Tourism Research 10(1): 9–33.
1983b   Tourism and Prostitution. Annals of Tourism Research 10(3):437–443.
1984   The Evolution of Tourist Arts. Annals of Tourism Research 11: 393–419.
1989   Tourism: the Sacred Journey. *In* Hosts and Guests: the Anthropology of Tourism (2nd ed.), V. Smith, ed., pp. 21–36. Philadelphia: University of Pennsylvania Press.
Graburn, Nelson H. H. and Jafar Jafari
1991   Introduction: Tourism Social Science. Annals of Tourism Research 18(1):1–11.
Gray, Jeffrey A.
1971   The Psychology of Fear and Stress. London: Weidenfeld and Nicolson.
Green, Nicholas
1990   The Spectacle of Nature: Landscape and Bourgeois Culture in Nineteenth-Century France. Manchester: Manchester University Press.
Greenblat, Cathy Stein and John H. Gagnon
1983   Tempoary Strangers: Travel and Tourism from a Sociological Perspective. Sociological Perspectives 26(1):89–110.
Gregory, Derek and John Urry, eds.
1985   Social Relations and Spatial Structures. London: Macmillan.
Griswold, W.
1994   Cultures and Societies in a Changing World. London: Pine Forge Press.
Grow, Gerald O.
1994   "Don't Hate Me Because I'm Beautiful: A Commercial in Context". *In* Images in Language, Media, and Mind, Roy F. Fox, ed., pp.170–180. Urbana, Illinois: National Council of Teachers of English.
Gunn, C.
1988   Vacationscapes: Designing Tourist Regions. New York: Van Nostrand Reinhold.
Gurvitch, Georges
1990   The Problem of Time. *In* The Sociology of Time, J. Hassard, ed., pp. 35–44. London: Macmillan.
Habermas, Jurgen
1984   The Theory of Communicative Action, vol. 1: Reason and the Rationalization of Society. Trans. by Thomas McCarthy. Cambridge: Polity Press.
1985   Modernity—an Incomplete Project. *In* Postmodern Culture, H. Foster, ed., pp. 3–15. London: Pluto.
1987   The Theory of Communicative Action, vol. 2: Lifeword and System: the Critique of Functionalist Reason. Trans. by Thomas McCarthy. Cambridge: Polity Press.
Hahn, Cynthia
1990   Local Sancta Souvenirs: Sealing the Pilgrim's Experience. *In* The Blessings of Pilgrimage, Robert Ousterhout, ed., pp. 85–96. Urbana: University of Illinois Press.

Hall, C. Michael
  1992   Sex Tourism in Southeast Asia. *In* Tourism and the Less Developed Countries, David Harrison, ed., pp.64–74. Chichester: Wiley.
  1994   Gender and Economic Interests in Tourism Prostitution: the Nature, Development and Implications of Sex Tourism in Southeast Asia. *In* Tourism: A Gender Analysis, Vivian Kinnaird and Derek Hall, eds., pp.142–163. Chichester: Wiley.
  1994   Tourism and Politics: Policy, Power and Place. New York: Wiley.
  1995   Tourism and Public Policy. London: Routledge.
Handler, R. and J. Linnekin
  1984   Tradition, Genuine or Spurious. Journal of American Folklore 97:273–290.
Handler, R. and W. Saxton
  1988   Dissimulation: Reflexivity, Narrative, and the Quest for Authenticity in "Living History". Cultural Anthropology 3(3):242–260.
Hanefors, M. and L. Larsson
  1993   Video Strategies Used by Tour Operators: What is really Communicated? Tourism Mangement 14(1):27–33.
Hannerz, Ulf
  1987   The World in Creolization. Africa 57(4):546–559.
  1990   Cosmopolitans and Locals in World Culture. Theory, Culture & Society 7:237–251.
Harkin, Michael
  1995   Modernist Anthropology and Tourism of the Authentic Annals of Tourism Research 22(3):650–670.
Harrison, David, ed.
  1992   Tourism and the Less Developed Countries. London: Bellhaven Press.
Harvey, David
  1985   Consciousness and the Urban Experience. Oxford: Basil Blackwell.
  1990   The Conditon of Postmodernity: an Enquiry into the Origins of Cultural Change. Oxford: Blackwell.
Hassard, John
  1990   Introduction: the Sociological Study of Time. *In* The Sociology of Time, John Hassard, ed., pp. 1–18.London: Macmillan.
Haukeland, Jan Vidar
  1990   Non-travelers: the Flip Side of Motivation. Annals of Tourism Research 17:172–184.
Heidegger, Martin
  1962   Being and Time. Oxford: Basil Blackwell.
  1978   Basic Writings, ed. D. Farrell Krell. London: Routledge & Kegan Paul.
Hern, Anthony
  1967   The Seaside Holiday: the History of the English Seaside Resort. London: The Cresset Press.
Hill, A. B. and R. E. Perkins
  1985   Towards a Model of Boredom. British Journal of Psychology 76:235–240.
Hirsch, Fred
  1977   Social Limits to Growth. London: Routledge & Kegan Paul.
Hitchcock, M. T. King and M. Parnwell, eds.
  1993   Tourism in South-east Asia. London: Routledge.
Hobsbawm, E. and T. Ranger
  1983   The Invention of Tradition. Cambridge: Cambridge University Press.

Hochschild, A. R.
1983 The Managed Heart: Commercialization of Human Feeling. London: University of California Press.

Hollinshead, Keith
1992 "White" Gaze, "Red" People—Shadow Visions. Leisure Studies 11: 43–64.
1996 Marketing and Metaphysical Realism. *In* Tourism and Indigenous People, R. Butler and T. Hinch, eds., pp. 308–348. London: International Thomson Business Press.
1997 Heritage Tourism under Post-modernity: Truth and the Past. *In* The Tourist Experience: A New Introduction, Chris Ryan, ed., pp. 170–193. London: Cassell.

Honderich, T., ed.
1995 The Oxford Companion to Philosophy. Oxford: Oxford University Press.

Horkheimer, M. and T. W. Adorno
1973 Dialectic of Enlightenment. Trans. J. Cumming. London: Allen Lane.

Horne, D.
1984 The Great Museum: the Representation of History. London: Pluto Press.

Hughes, George
1992 Tourism and the Geographical Imagination. Leisure Studies 11(1):31–42.
1995 Authenticity in Tourism. Annals of Tourism Research 22(4):781–803.

Huizinga, J.
1949 Homo Ludens: a Study of the Play-element in Culture. London: Routledge & Kegan Paul.

Hultsman, John
1995 Just Tourism: an Ethical Framework. Annals of Tourism Research 22(3):553–567.

Hummon, David M.
1988 Tourist Worlds: Tourist Advertising, Ritual, and American Culture. Sociological Quarterly 29(2):179–202.

Hunt, John D.
1975 Image as a Factor in Tourism Development. Journal of Travel Research 13(3):1–7.

Husbands, W.
1981 Centres, Peripheries, Tourism and Socio-spatial Development. Ontario Geography 17:37–59.

Iso-Ahola, S. E.
1983 Toward a Social Psychology of Recreational Travel. Leisure Studies 2:45–56.

IUOTO
1963 The United Nation's Conference on International Travel and Tourism. Geneva: International Union of Official Travel Organisations.

Jafari, Jafar.
1977 Editors Page. Annals of Tourism Research 5(special number):6–11.

James, W.
1890 The Principles of Psychology. New York: Holt.

Jaques, Elliott
1990 The Enigma of Time. *In* The Sociology of Time, John Hassard, ed., pp. 21–34. London: Macmillan.

Jasen, Patricia
   1991   Romanticism, Modernity, and the Evolution of Tourism on the Niagara
     Frontier, 1790–1850, Canadian Historical Review 72(3):283–318.
Johnson, Peter and Barry Thomas
   1992   The Analysis of Choice and Demand in Tourism. *In* Choice and
     Demand in Tourism. Peter Johnson and Barry Thomas, eds., pp. 1–12.
     London: Mansell.
Jokinen, Eeva and Soile Veijola
   1997   The Disoriented Tourist: the Figuration of the Tourist in Contemporary
     Cultural Critique. *In* Touring Cultures: Transformations of Travel
     and Theory, Chris Rojek and John Urry, eds., pp. 23–51. London:
     Routledge.
Jules-Rosette, Bennetta
   1984   The Messages of Tourist Art: An African Semiotic System in
     Comparative Perspective. New York: Plenum Press.
Kabbani, Rana
   1986   Europe's Myths of Orient. Bloomington: Indiana University Press.
Kaplan, M.
   1960   Leisure in America: A Social Inquiry. New York: John Wiley.
Karl, Herb
   1994   The Image is not the Thing. *In* Images in Language, Media, and Mind,
     Roy F. Fox, ed., pp. 193–203. Urbana, Illinois: National Council of Teachers
     of English.
Kern, S.
   1983   The Culture of Time and Space: 1880–1918. Cambridge: Harvard
     University Press.
King, A. D., ed.
   1991   Culture,   Globalization   and   the   World-system: Contemporary
     Conditions   for   the   Representation   of   Identity.   Binghamton: State
     University of New York; London: Macmillan.
Kinnaird, Vivian and Derek Hall, eds.
   1994   Tourism: A Gender Analysis. Chichester: Wiley.
Knebel, H. J.
   1960   Soziologische Strukturwandlungen im modernen Tourismus. Stuttgart:
     Enke.
Kolb, D.
   1986   The Critique of Pure Modernity: Hegel, Heidegger and After. London:
     The University of Chicago Press.
Krippendorf, Jost
   1987   The Holiday Maker: Understanding the Impact of Leisure and Travel.
     Trans. by Vera Andrassy. London: Heinemann.
Kuhn, Thomas S.
   1962   The Structure of Scientific Revolutions. Chicago: University of Chicago
     Press.
Kumar, Krishan
   1995   From Post-industrial to Post-modern Society: New Theories of the
     Contemporary World. Oxford: Blackwell.
Lakoff, G. and M. Johnson
   1980   Metaphors We Live by. Chicago: Chicago University Press.
Lanfant, M.-F.
   1980   Introduction: Tourism in the Process of Internationalization.
     International Social Science Journal 32(1):14–43.

1989   International Tourism Resists the Crisis. *In* Leisure and life-style, A. Olszewska and K. Roberts, eds., pp. 178–193. London: Sage.

1993   Methodological and Conceptual Issues Raised by the Study of International Tourism: A Test of Sociology. *In* Tourism Research: Critiques and Challenges, Douglas G. Pearce and Richard W. Butler, eds., pp. 70–87. London: Routledge.

1995a   Introduction. *In* International Tourism: Identity and Change, M.-F. Lanfant, J. B. Allcock and E. M. Bruner, eds., pp. 1–23. London: Sage.

1995b   International Tourism, Internationalization and the Challenge to Identity. *In* International Tourism: Identity and Change, M.-F. Lanfant, J. B. Allcock and E. M. Bruner, eds., pp.24–43. London: Sage.

Lanfant, M.-F. , J. B. Allcock and E. M. Bruner, eds.
1995   International Tourism: Identity and Change. London: Sage.

Langman, Lauren
1992   Neon Cages: Shopping for Subjectivity. *In* Lifestyle Shopping: the Subject of Consumption, Rob Shields, ed., pp. 40–82. London: Routledge.

Lasch, C.
1979   The Culture of Narcissism. London: AACUS.

Lash, Scott
1993   Reflexive Modernization: the Aesthetic Dimension. Theory, Culture & Society 10(1):1–23.

Lash, Scott and John Urry
1994   Economy of Signs and Space. London: Sage.

Laskey, Henry A., Bruce Seaton and J. A. F. Nicholls
1994   Effects of Strategy and Pictures in Travel Agency Advertising. Journal of Travel Research 32(4):13–19.

Laxson, J. D.
1991   How "We" See "Them": Tourism and Native Americans. Annals of Tourism Research 18(3):365–391.

Lea, John
1988   Tourism and Development in the Third World. London: Routledge.
1993   Tourism Development Ethics in the Third World. Annals of Tourism Research 20:701–715.

Lee, Martyn
1993   Consumer Culture Reborn: The Cultural Politics of Consumption. London: Routledge.

Lee, Tae-Hee, and John Crompton
1992   Measuring Novelty Seeking in Tourism. Annals of Tourism Research 19(4):732–751.

Lee, Wendy
1991   Prostitution and Tourism in Southeast Asia. *In* Working Women: International Perspectives on Labour and Gender Ideology, Nanneke Redclift and M. Thea Sinclair, eds., pp. 79–103. London: Routledge.

Lefebvre, H.
1991   The Production of Space. Oxford: Blackwell.

Leheny, David
1995   A Political Economy of Asia Sex Tourism. Annals of Tourism Research 22(2):367–84.

Leidner, Robin
1993   Fast Food, Fast Talk: Service Work and the Routinisation of Everyday Life. London: University of California Press.

Leiper, Neil
   1979   The Framework of Tourism: towards a Definition of Tourism, Tourist and the Tourism Industry. Annals of Tourism Research 6(4): 390–407.
   1983   An Etymology of "Tourism". Annals of Tourism Research 10(2):277–280.
Leiss, W., S. Kline and S. Jhally
   1997   Social Communication in Advertising (2nd ed.). London: Routledge.
Leslie, D.
   1994   Sustainable Tourism or Developing Sustainable Approaches to Lifestyle? World Leisure & Recreation 36(3):30–36.
Lett, James W.
   1983   Ludic and Liminoid Aspects of Charter Yacht Tourism in the Caribbean. Annals of Tourism Research 10(1):35–56.
Levitt, Eugene Elmer
   1968   The Psychology of Anxiety. London: Staples Press.
Lewis, J. D. and A. J. Weigart
   1990   The Structures and Meanings of Social-Time. *In* The Sociology of Time, John Hassard, ed., pp. 77–101. London: Macmillan.
Littrell, Mary Ann
   1990   Symbolic Significance of Textile Crafts for Tourists. Annals of Tourism Research 17:228–245.
Littrell, M. A., L. F. Anderson and P. J. Brown
   1993   What Makes A Craft Souvenir Authentic? Annals of Tourism Research 20(1):197–215.
Littrell, M. A., S. Baizerman, R. Kean, S. Gahring, S. Niemeyer, R. Reilly, and J. A. Stout
   1994   Souvenirs and Tourism Styles. Journal of Travel Research 33(1):3–11.
Lockwood, David
   1964   Social Integration and System Integration. *In* Explorations in Social Change, G. K. Zollschan and W. Hirsch, eds., pp. 244–257. London: Routledge.
Lollar, Sam A. and Carlton Van Doren
   1991   U. S. Tourist Destinations: A History of Desirability. Annals of Tourism Research 18:622–638.
Lowenthal, David
   1985   The Past is a Foreign Country. Cambridge: Cambridge University Press.
Lowy, Michael
   1987   The Romantic and the Marxist Critique of Modern Civilization. Theory and Society 16(6):891–904.
Ludz, Peter C.
   1973   Alienation as a Concept in the Social Sciences Current Sociology 21(1):9–103.
Lury, Celia
   1996   Consumer Culture. Cambridge: Polity Press.
Lyman, Stanford M. and Marvin B. Scott
   1970   A Sociology of the Absurd. California: Goodyear
Lyon, D.
   1994   Postmodernity. Buckingham: Open University Press.
Lyotard, J.-F.
   1984   The Postmodern Condition. Manchester: Manchester University Press.

Macbeth, Jim
    1988   Ocean Cruising. *In* Optimal Experience: Psychological Studies of Flow in Consciousness. M. Csikszentmihalyi and I. S. Csikszentmihalyi, eds., pp. 214–231. Cambridge: Cambridge University Press.

MacCannell, Dean
    1973   Staged Authenticity: Arrangements of Social Space in Tourist Settings. American Journal of Sociology 79(3):589–603.
    1976   The Tourist: a New Theory of the Leisure Class. New York: Schocken.
    1989   Introduction to the 1989 Edition. *In* The Tourist (2nd ed.) New York: Schocken.
    1992   Empty Meeting Grounds: the Tourist Papers. London: Routledge.

Macdonald, Sharon
    1997   A People's Story: Heritage, Identity and Authenticity. *In* Touring Cultures: Transformations of Travel and Theory. Chris Rojek and John Urry, eds., pp. 155–175. London: Routledge.

Machlis, Gary E. and William R. Burch
    1983   Relations between Strangers: Cycles of Structure and Meaning in Tourist Systems. Sociological Review 31(4):666–692.

MacKay, Kelly J. and Daniel R. Fesenmaier
    1997   Pictorial Element of Destination in Image Formation. Annals of Tourism Research 24(3):537–565.

Maffesoli, M.
    1996   The Time of the Tribes. Trans. by Don Smith. London: Sage.

Malinowski, B.
    1990   Time-reckoning in the Trobriands. *In* The Sociology of Time, John Hassard, ed., pp. 203–218. London: Macmillan.

Marcuse, H.
    1955   Eros and Civilisation. London: The Beacon Press.

Marshment, Margaret
    1997   Gender Takes a Holiday: Representation in Holiday Brochures. *In* Gender, Work and Tourism, M. Thea Sinclair, ed., pp. 16–34. London: Routledge.

Marx, Karl
    1954   Capital, vols. 1–3. London: Lawrence & Wishart.
    1977   Economic and Philosophic Manuscripts of 1844. London: Lawrence & Wishart.

Mathieson, Alister and Geoffrey Wall
    1982   Tourism: Economic, Physical and Social impacts. London: Longman.

McCrone, D., A. Morris and R. Keily
    1995   Scotland-the Brand: the Making of Scottish Heritage. Edinburgh: Edinburgh University Press.

McLuhan, Marshall
    1964   Understanding Media. London: Routledge.

Mead, George Herbert
    1934   Mind, Self, and Society. Chicago: The University of Chicago Press.

Mellinger, Wayne Martin
    1994   Toward a Critical Analysis of Tourism Representation. Annals of Tourism Research 21(4):756–779.

Mergen, Bernard
    1986   Travel as Play. *In* Cultural Dimensions of Play, Games, and Sport, Bernard Mergen, ed., pp.103–111. Champaign, Illinois: Human Kinetics Publishers.

Merton, Robert K.
  1976 Sociological Ambivalence and Other Essays. New York: Free Press.
Michael, M. A., ed.
  1950a Traveler's Quest: Original Contributions towards a Philosophy of Travel. London: William Hodge.
Michael, M. A.
  1950b What is Travel? An Introduction. *In* Traveler's Quest: Original Contributions towards a Philosophy of Travel, M. A. Michael, ed., pp.1–19. London: William Hodge.
Mikulas, William L. and Stephen J. Vodanovich
  1993 The Essence of Boredom. The Psychological Record 43:3–12.
Mills, C. Wright
  1951 White Collar: the American Middle Classes. Oxford: Oxford University Press.
  1959 The Sociological Imagination. Oxford: Oxford University Press.
Mitchell, Richard G.
  1983 Mountain Experience: the Psychology and Sociology of Adventure. Chicago: University of Chicago Press.
  1988 Sociological Implications of the Flow Experience. In Optimal Experience: Psychological Studies of Flow in Consciousness, M. Csikszentimihalyi and I. S. Csikszentmihalyi, eds., pp. 36–59. Cambridge: Cambridge University Press.
Mitford, N.
  1959 The Tourist. Encounter 13(4):3–7.
Moeran, Brian
  1983 The Language of Japanese Tourism. Annals of Tourism Research 10(1):93–108.
Moog, Carol
  1994 Ad Images and the Stunting of Sexuality. *In* Images in Language, Media, and Mind, Roy F. Fox, ed., pp. 152–169. Urbana, Illinois: National Council of Teachers of English.
Moore, W. E.
  1963 Man, Time and Society. New York: Wiley.
Morley, David and Kevin Robins
  1995 Spaces of Identity: Global Media, Electronic Landscapes and Cultural Boundaries. London: Routledge.
Moscardo, G. M., and P. L. Pearce
  1986 Historic Theme Parks: an Australian Experience in Authenticity. Annals of Tourism Research 13(3):467–479.
Mukerjee, Radhakamal
  1990 Time, Technics and Society. *In* The Sociology of Time, J. Hassard, ed., pp. 47–55. London: Macmillan.
Mumford, Lewis
  1934 Technics and Civilization. New York: Harcourt, Brace & World.
  1966 Technics and the Nature of Man. Technology and Culture 7(3): 303–317.
Munt, Ian
  1994 The "Other" Postmodern Tourism: Culture, Travel and New Middle Classes. Theory, Culture & Society 11:101–123.
Murphy, Peter
  1985 Tourism: a Community Approach. London: Routledge.

Murphy, Peter
   1994   Tourism and Sustainable Development. *In* Global Tourism: the Next
   Decade, William F. Theobald, ed., pp. 274–290. London: Butterworth-
   Heinemann.
Nasar, Herbert, ed.
   1988   Environmental Aesthetics. Cambridge: Cambridge University Press.
Nash, Dennison
   1981   Tourism as an Anthropological Subject. Current Anthropology
   22(5):461–481.
   1989   Tourism as a Form of Imperialism. *In* Hosts and Guests: the
   Anthropology of Tourism (2nd ed.), V. Smith, ed., pp. 37–52.
   Philadelphia: University of Pennsylvania Press.
   1995   An Exploration of Tourism as Superstructure. *In* Change in Tourism:
   People, Places, Processes, Richard Butler and Douglas Pearce, eds., pp. 30–
   46. London: Routledge.
Nelson, J. G., R. Butler and G. Wall, eds.
   1993   Tourism and Sustainable Development: Monitoring, Planning,
   Managing. Waterloo: Heritage Resources Center and Department of
   Geography, University of Waterloo.
Netton, Ian Richard
   1990   The Mysteries of Islam. *In* Exoticism in the Enlightenment, G. S.
   Rousseau and Roy Porter, eds., pp. 23–45. Manchester: Manchester
   University Press.
Neumann, Mark
   1992   The Trail through Experience: Finding Self in the Recollection
   of Travel. *In* Investigating Subjectivity: Research on Lived Experience,
   Carolyn Ellis and Michael G. Flaherty, eds., pp. 176–201. London:
   Sage.
Norval, A. J.
   1936   The Tourist Industry: a National and International Survey. London:
   Pitman.
Nunez, Th. A.
   1963   Tourism, Tradition and Acculturation: Weekendismo in a Mexican
   Village. Ethnology 2(3):347–352.
Ogilvie, F. W.
   1933   The Tourist Movement: an Economic Study. London: Staples.
O'Hanlon, James F.
   1981   Boredom: Practical Consequences and Theory. Acta Psychologica
   49(1):53–82.
O'Neill, Edward
   1972   Rape of the American Virgins. New York: Praeger.
Ousby, Ian
   1990   The Englishman's England: Taste, Travel and the Rise of Tourism.
   Cambridge: Cambridge University Press.
Pearce, Douglas
   1989   Tourism Development (2nd ed.). Harlow, Essex: Longman Scientific &
   Technical.
Pearce, Philip L.
   1982   The Social Psychology of Tourist Behaviour. Oxford: Pergamon.
   1988   The Ulysses Factor: Evaluating Visitors in Tourist Settings. New York:
   Springer-Verlag.

Pearce, P. L. and G. M. Moscardo
  1985   The Relationship between Travelers' Career Levels and the Concept of
    Authenticity. Australian Journal of Psychology 37:157–174.
  1986   The Concept of Authenticity in Tourist Experiences. Australian and
    New Zealand Journal of Sociology 22(1):121–132.
Perkins, R. E. and A. B. Hill
  1985   Cognitive and Affective Aspects of Boredom. British Journal of
    Psychology 76:221–234.
Pettman, Jan Jindy
  1997   Body Politics: International Sex Tourism. Third World Quarterly
    18(1):93–108.
Picard, Michel
  1995   Cultural Heritage and Tourist Capital: Cultural Tourism in Bali. *In*
    International Tourism: Identity and Change, M.-F. Lanfant, J. B. Allcock
    and E. D. Bruner, eds., pp. 44–66. London: Sage.
Pimlott, J. A. R.
  1976   The Englishman's Holiday: A Social History. Brighton: The Harvester
    Press (first published in London: Faber & Faber, 1947).
Plog, Stanley C.
  1994   Leisure Travel: an Extraordinary Industry Faces Superordinary
    Problems. *In* Global Tourism: the Next Decade, W. Theobald, ed., pp.40–
    54. London: Butterworth-Heinemann.
Poiesz, Theo B. C.
  1989   The Image Concept: Its Place in Consumer Psychology. Journal of
    Economic Psychology 10:457–472.
Poon, A.
  1993   Tourism, Technology and Competitive Strategies. Wallingford: CAB
    International.
Popper, K. R.
  1969   Conjectures and Refutation. London: Routledge & Kegan Paul.
Porter, Roy
  1990   The Exotic as Erotic: Captain Cook at Tahiti. *In* Exoticism in the
    Enlightenment, G. S. Rousseau and Roy Porter, eds., pp. 117–144.
    Manchester: Manchester University Press.
Poster, Mark
  1990   The Mode of Information. Cambridge: Polity Press.
Postman, Neil
  1985   Amusing Ourselves to Death: Public Discourse in the Age of Show
    Business. London: Heinemann.
Poulantzas, N.
  1978   State, Power, Socialism. London: Verso.
Prentice, Richard
  1997   Cultural and Landscape Tourism: Facilitating Meaning. *In* Tourism,
    Development and Growth: the Challenge of Sustainability, Salah Wahab
    and John J. Pigram, eds., pp. 209–236. London: Routledge.
Przeclawski, Krzysztof
  1993   Tourism as the Subject of Interdisciplinary Research. *In* Tourism
    Research, Douglas G. Pearce and Richard W. Butler, eds., pp. 9–20.
    London: Routledge.
Pucci, Suzanne Rodin
  1990   The Discrete Charms of the Exotic: Fictions of the Harem in
    Eighteenth-century France. *In* Exoticism in the Enlightenment, G. S.
    Rousseau and Roy Porter, eds. Manchester: Manchester University Press.

Redfoot, D.
  1984   Touristic Authenticity, Touristic Angst, and Modern Reality. Qualitative sociology 7(4):291–309.
Reimer, Gwen Dianne
  1990   Packaging Dreams: Canadian Tour Operators at Work. Annals of Tourism Research 17:501–512.
Reisinger, Y.
  1994   Social Contact between Tourists and Hosts of Different Cultural Backgrounds. *In* Tourism: the State of the Art, A. V. Seaton, ed., pp. 743–754. Chichester: Wiley.
Report of the Committee on Holidays with Pay
  1938   Presented by the Minister of Labour to Parliament by Command of His Majesty, April, 1938. London: HMSO.
Riasanovsky, Nicholas V.
  1992   The Emergence of Romanticism. Oxford: Oxford University Press.
Richter, Linda K.
  1989   The Politics of Tourism in Asia. Honolulu: University of Hawaii Press.
Ritzer, George
  1992   Sociological Theory (3rd ed.). New York: McGraw-Hill.
  1996   The McDonaldiszation of society (revised ed.) Thousand Oaks, California: Pine Forge Press.
Ritzer, G. and A. Liska
  1997   "McDisneyization" and "Post-Tourism": Complementary Perspectives on Contemporary Tourism. *In* Touring Cultures, Chris Rojek and John Urry, eds., pp. 96–109. London: Routledge.
Robertson, Roland
  1992   Globalisation: Social Theory and Global Culture. London: Sage.
Roche, Maurice
  1973   Phenomenology, Language and the Social Science. London: Routledge and Kegan Paul.
  1990   Time and Unemployment. Human Studies 13:73–96.
  1992   Mega-events and Micro-Modernization: on the Sociology of the New Urban Tourism. British Journal of Sociology 43(4):563–600.
  1994   Mega-events and Urban Policy. Annals of Tourism Research 21(1):1–19.
Rojek, Chris
  1993   Ways of Escape: Modern Transformations in Leisure and Travel. London: Macmillan.
Rojek, Chris
  1995   Decentering Leisure: Rethinking Leisure Theory. London: Sage.
Rojek, Chris
  1997   Indexing, Dragging and the Social Construction of Tourist Sights. *In* Touring Cultures: Transformations of Travel and Theory, Chris Rojek and John Urry, eds., pp. 52–74. London: Routledge.
Rojek, Chris and John Urry, eds.
  1997   Touring Cultures: Transformations of Travel and Theory. London: Routledge.
Rosaldo, Renato
  1989   Imperial Nostalgia. Representations 26 (Spring):107–122.
Ross, Glenn F.
  1994   The Psychology of Tourism. Melbourne: Hospitality Press.
Rousseau, G. S. and Roy Porter, eds.
  1990a   Exoticism in the Enlightenment. Manchester: Manchester University Press.

Rousseau, G. S. and Roy Porter
   1990b   Introduction. *In* Exoticism in the Enlightenment, G. S. Rousseau and
      Roy Porter, eds., pp. 1–22. Manchester: Manchester University Press.
Russell, Bertrand
   1979   History of Western Philosophy. London: Book Club Associates.
Russell, James A.
   1988   Affective Appraisal of Environments. *In* Environmental Aesthetics, H.
      Nasar, ed.) pp.120–132. Cambridge: Cambridge University Press.
Rutledge, Kay Ellen
   1994   Analyzing Visual Persuasion: The Art of Duck Hunting. *In* Images in
      Language, Media, and Mind, Roy F. Fox, ed., pp. 204–218. Urbana, Illinois:
      National Council of Teachers of English.
Ryan, Chris
   1991   Recreational Tourism: a Social Science Perspective. London:
      Routledge.
Ryan, Chris, ed.
   1997a   The Tourist Experience: A New Introduction. London: Cassell.
Ryan, Chris
   1997b   "The Time of Our Lives" or Time for Our Lives: An Examination of
      Time in Holidaying. *In* The Tourist Experience: A New Introduction, Chris
      Ryan, ed., pp. 194–205. London: Cassell.
Sabel, Charles
   1982   Work and Politics: the Division of Labour in Industry. Cambridge:
      Cambridge University Press.
Sack, Robert David
   1986   Human Territoriality: Its Theory and History. Cambridge: Cambridge
      University Press.
   1992   Place, Modernity, and the Consumer's World: a Relational Framework
      for Geographical Analysis. London: Johns Hopkins University Press.
Said, Edward
   1978   Orientalism. London: Routledge.
Salamone, F. A.
   1997   Authenticity in Tourism: The San Angel Inns. Annals of Tourism
      Research 24(2):305–321.
Saram, P. A.
   1983   Tourist Regions: Some Sociological Observations. California
      Sociologist 6(2):89–103.
Saunders, Peter
   1985   Space, the City and Urban Sociology. *In* Social Relations and Spatial
      Structures. Derek Gregory and John Urry, eds., pp. 67–89. London:
      Macmillan.
   1989   Space, Urbanism, and Created Environment. *In* Social Theory of
      Modern Societies: Anthony Giddens and His Critics, David Held and John
      B. Thompson, eds., pp. 215–234. Cambridge: Cambridge University
      Press.
Sayer, Andrew
   1985   The Difference that Space Makes. *In* Social Relations and Spatial
      Structures, Derek Gregory and John Urry, eds., pp. 49–66. London:
      Macmillan.
Schivelbusch, Wolfgang
   1986   The Railway Journey: the Industrialization of Time and Space in the
      19th Century. Leamington Spa: Berg.

Schmidt, Catherine J.
  1979   The Guide Tour: Insulated Adventure. Urban Life 7(4):441–467.
Schudson, M.
  1979   Review Essay: on Tourism and Modern Culture. American Journal of Sociology 84(5):1249–1258.
Schutz, Alfred, and Thomas Luckmann
  1974   The Structures of the Life-World. London: Heinemann.
Schwandt, T. A.
  1994   Constructivist, Interpretivist Approaches to Human Inquiry. *In* Handbook of Qualitative Research, Norman K. Denzin and Yvonna S. Lincoln, eds., pp. 118–137. London: Sage.
Seamon, David
  1979   A Geography of the Lifeworld: Movement, Rest and Encounter. London: Croom Helm.
Seligman, Adrian
  1950   The Call of the Sea. *In* Travelers' Quest: Original Contributions towards a Philosophy of Travel, M. A. Michael, ed., pp. 74–89. London: William Hodge.
Selwyn, Tom
  1993   Peter Pan in South-East Asia: Views from the Brochures. *In* Tourism in South-East Asia, M. Hitchcock, V. King and M. Parnwell, eds., pp. 117–137. London: Routledge.
  1996a   Introduction. *In* The Tourist Image: Myths and Myth Making in Tourism, Tom Selwyn, ed., pp. 1–32. Chichester: Wiley.
Selwyn, Tom, ed.
  1996b   The Tourist Image: Myths and Myth Making in Tourism. Chichester: Wiley.
Shanks, M.
  1992   Experiencing the Past. London: Routledge.
Sharpley, Richard
  1994   Tourism, Tourists & Society. Huntingdon: Elm Publications.
Shaw, Gareth and Allan M. Williams
  1994   Critical Issues in Tourism: a Geographical Perspective. Oxford: Blackwell.
Shenhav-Keller, Shelly
  1993   The Israeli Souvenir: Its Text and Context. Annals of Tourism Research 20(1):182–196.
Shields, Rob
  1991   Places on the Margin: Alternative Geography of Modernity. London: Routledge.
  1992   A Truant Proximity: Presence and Absence in the Space of Modernity. Environment and Planning D: Society and Space 10 (2):181–198.
Short, John Rennie
  1991   Imagined Country: Environment, Culture and Society. London: Routledge.
Silver, Ira
  1993   Marketing Authenticity in Third World Countries. Annals of Tourism Research 20(2):302–318.
Simmel, Georg
  1950a   The Stranger. *In* The Sociology of George Simmel, trans. and ed. by Kurt H. Wolff, pp. 402–424. Glencoe, Illinois: Free Press.
  1950b   The Metropolis and Mental Life. *In* The Sociology of Georg Simmel, trans. and ed. by Kurt H. Wolff. Glencoe, Illinois: Free Press.

Simmel, Georg
   1990   Philosophy of Money (2nd enlarged ed.), trans. by T. Bottomore and D.
   Frisby. London: Routledge.
Slater, Don
   1997   Consumer Culture and Modernity. Cambridge: Polity Press.
Smith, Adam
   1910   The Wealth of Nations. London: Dent.
Smith, Anthony D.
   1981   The Ethnic Revival. Cambridge: Cambridge University Press.
Smith, Stephen L. J.
   1988   Defining Tourism: a Supply-side View. Annals of Tourism Research
   15:179–190.
Smith, V. and W. Eadington
   1992   Tourism Alternative: Potentials and Problems in the Development of
   Tourism. Philadelphia: University of Pennsylvania Press.
Smith, Valene L., ed.
   1977   Hosts and Guests: the Anthropology of Tourism. Philadelphia:
   University of Pennsylvania Press.
   1989   Hosts and Guests: the Anthropology of Tourism (2nd ed.) Philadelphia:
   University of Pennsylvania Press.
Soja, Edward W.
   1989   Postmodern Geographies: the Reassertion of Space in Critical Social
   Theory. London: Verso.
Sontag, S.
   1979   On Photography. Harmondsworth: Penguin.
Sorokin, Pitirim and Robert Merton
   1990   Social Time: a Methodological and Functional Analysis. *In* The
   Sociology of Time, John Hassard, ed., pp. 56–66. London: Macmillan.
Spielberger, Charles D., ed.
   1972   Anxiety: Current Trends in Theory and Research, vols. 1 and 2.
   London: Academic Press.
Spielberger, Charles D. and Laura M. Starr
   1994   Curiosity and Exploratory Behavior. *In* Motivation: Theory and
   Research, Harold F. O'Neil, Jr. and Michael Drillings, eds. Hove, UK:
   Lawrence Erlbaum Associates.
Spooner, B.
   1986   Weavers and Dealers: the Authenticity of an Oriental Carpet. *In* The
   social life of things, A. Appadurai, ed., pp. 195–235. Cambridge: Cambridge
   University Press.
Squire, Shelagh J.
   1988   Wordsworth and Lake District Tourism: Romantic Reshaping of
   Landscape. The Canadian Geographer 32(3):237–247.
Stagl, Justin
   1995   A History of Curiosity: The Theory of Travel 1550–1800. Chur:
   Harwood Academic Publishers.
Starkey, Kern
   1988   Time and Work Organisation: a Theoritical and Empirical Analysis.
   In The rhythms of society, M. Young and T. Schuller, eds. London:
   Routledge.
Stephen, C.
   1990   The Search for Authenticity: Review Essay of Dean MacCannell, The
   Tourist. Berkeley Journal of Sociology: a Critical Review 35:151–156.

Stewart, Susan
    1984   On Longing: Narratives of the Miniature, the Gigantic, the Souvenir,
    the Collection. Baltimore: Johns Hopkins Unversity Press.
Stubbs, Michael
    1983   Discourse Analysis: The Sociolinguistic Analysis of Natural Language.
    Oxford: Basil Blackwell.
Swinglehurst, Edmund
    1974   The Romantic Journey: the Story of Thomas Cook and Victorian Travel.
    London: Pica Editions.
Taylor, Charles
    1989   Sources of the Self: the Making of the Modern Identity. Cambridge,
    Massachusetts: Harvard University Press.
    1991   The Ethics of Authenticity. Cambridge, Massachusetts: Harvard
    University Press.
Telisman-Kosuta, Neda
    1989   Tourist Destination Image. *In* Tourism Marketing and Management
    Handbook, Stephen F. Witt and Luiz Moutinho, eds., pp. 557–61.
    London: Prentice Hall.
Tester, Keith
    1993   The Life and Times of Post-modernity. London: Routledge.
Theobald, William F.
    1994a   The Context, Meaning and Scope of Tourism. *In* Global Tourism: the
    Next Decade, F. William Theobald, ed., pp. 3–19. Oxford: Butterworth-
    Heinemann.
Theobald, William F., ed.
    1994b   Global Tourism: the Next Decade. Oxford: Butterworth-Heinemann.
Thompson, E. P.
    1967   Time, Work-discipline and Industrial Capitalism. Past and Present
    38:56–97.
Thompson, Michael, Richard Ellis and Aaron Wildavsky
    1990   Cultural Theory. Boulder: Westview Press.
Thrift, Nigel
    1990   The Making of a Capitalist Time Consciousness. *In* The Sociology of
    Time, John Hassard, ed., pp. 105–129. London: Macmillan.
Thurot, Jean Maurice and Gaétane Thurot
    1983   The Ideology of Class and Tourism: Confronting the Discourse of
    Advertising. Annals of Tourism Research 10(1):173–189.
Todorov, Tzvetan
    1993   On Human Diversity: Nationalism, Racism, and Exoticism in French
    Thought trans. by Catherine Porter. Cambridge, Massachusetts: Harvard
    University Press.
Tönnies, Ferdinand
    1955   Community and Association. London: Routledge & Kegan Paul.
Towner, John
    1985   The Grand Tour: A Key Phase in the History of Tourism. Annals of
    Tourism Research 12(3):297–333.
Trilling, L.
    1972   Sincerity and Authenticity. London: Oxford University Press.
Tuan, Yi-Fu
    1974   Topophilia: a Study of Environmental Perception, Attitudes, and
    Values. New York: Columbia University Press.
    1977   Space and Place: the Perspective of Experience. London: University of
    Minnesota Press.

Turner, B.
   1987   A Note on Nostalgia. Theory, Culture & Society 4:147–156.
Turner, C. and P. Manning
   1988   Placing Authenticity—on Being a Tourist: A Reply to Pearce and Moscardo. Australia and New Zealand Journal of Sociology 24:136–139.
Turner, Louis and John Ash
   1975   The Golden Hordes: International Tourism and the Pleasure Periphery. London: Constable.
Turner, Victor
   1969   The Ritual Process. Chicago: Aldine.
   1973   The Center Out There: Pilgrims' Goal. History of Religion 12(3):191–230.
   1974   Dramas, Fields, and Metaphors. Ithaca, New York: Cornell University Press.
Turner, V. and E. Turner
   1978   Image and Pilgrimage in Christian Culture. Oxford: Basil Blackwell.
UNESCO
   1976   The Effects of Tourism on Socio-cultural Values. Annals of Tourism Research 4(1):74–105.
Urbain, Jean-Didier
   1989   The Tourist Adventure and His Images. Annals of Tourism Research 16(1):106–118.
Urry, John
   1985   Social Relations, Space and Time. *In* Social Relations and Spatial Structures, Derek Gregory and John Urry, eds., pp. 20–48. London: Macmillan.
   1988   Cultural Change and Contemporary Holiday-making, Theory. Culture & Society 5:35–55.
   1990a   The Tourist Gaze: Leisure and Travel in Contemporary Societies. London: Sage.
   1990b   The "Consumption" of Tourism. Sociology 24(1):23–35.
   1991a   The Sociology of Tourism. *In* Progress in Tourism, Recreation and Hospitality Management, vol. 3, C. P. Cooper, ed., pp. 48–57. London: Behaven Press.
   1991b   Time and Space in Giddens' Social Theory. *In* Giddens' Theory of Structuration: a Critical Appreciation, Christopher G. A. Bryant and David Jary, eds., pp. 160–75. London: Routledge.
   1992   The Tourist Gaze and the "Environment". Theory, Culture & Society 9:1–26.
   1994a   Cultural Change and Contemporary Tourism. Leisure Studies 13:233–38.
   1994b   Europe, Tourism and the Nation-state. *In* Progress in Tourism, Recreation and Hospitality Management, vol. 5, C. P. Cooper and A. Lockwood, eds., pp. 89–98. Chichester: Wiley.
   1995   Consuming Places. London: Routledge.
Uzzell, D. L.
   1989   The Hot Interpretation of War and Conflict. *In* Heritage Interpretation, vol. 1: the Natural and Built Environment, D. L. Uzzell, ed., pp. 33–47. London: Belhaven Press.
Van den Abbeele, Georges
   1980   Sightseers: the Tourist as Theorist. Diacritics 10:2–14.

Vanhove, Norbert
  1997 Mass Tourism: Benefits and Costs. *In* Tourism, Development and Growth, Salah Wahab and John J. Pigram, eds., pp.50–77. London: Routledge.
Veal, A. J.
  1993 The Concept of Lifestyle: A Review. Leisure Studies, 12(4):233–252.
Veblen, Thorstein
  1925 The Theory of the Leisure Class. London: Allen & Unwin.
Veijola, S. and E. Jokinen
  1994 The Body in Tourism. Theory, Culture & Society 11:125–151.
Vester, H.-G.
  1987 Adventure as a Form of Leisure. Leisure Studies 6(3):237–249.
Voase, Richard
  1995 Tourism: the Human Perspective. London: Hodder & Stoughton.
von Wiese, L.
  1930 Fremdenverkehr als zwischenmenschliche Beziehung. Arch. Fremdenverkehr, 1(1).
Vukonić, Boris
  1996 Tourism and Religion, trans. by Sanja Matešic. Oxford: Elsevier Science.
Wachtel, Paul L.
  1983 The Poverty of Affluence: a Psychological Portrait of the American Ways of Life. New York: Free Press.
Wagner, Peter
  1994 A Sociology of Modernity: Liberty and Discipline. London: Routledge.
Walker, C. R. and R. Guest
  1952 Man on the Assembly Line. Cambridge, Massachussetts: Harvard University Press.
Wallerstein, Immanuel
  1979 The Rise and Future Demise of the World Capitalist System: Concepts for Comparative Analysis. *In* The Capitalist World Economy, I. Wallerstein. Cambridge: Cambridge University Press.
Wang, Ning
  1996 Logos-modernity, Eros-modernity, and Leisure. Leisure Studies 15(2):121–135.
  1997a Vernacular House as Attraction: an Illustration from Hutong Tourism in Beijing. Tourism Management 18(8):573–580.
  1997b Modernity, Ambivalence, and Tourism: Towards A Sociology of Tourism. PhD Dissertation in Sociology. University of Sheffield, UK.
Waters, Malcolm
  1995 Globalisation. London: Routledge.
Watson, G. Llewellyn and Joseph P. Kopachevsky
  1994 Interpretation of Tourism as Commodity. Annals of Tourism Research 21(3):643–660.
Weber, Max
  1970 The Protestant Ethic and the Spirit of Capitalism. London: Unwin University Books.
  1978 Economy and Society, vols. 1 and 2, edited by Guenther Roth and Claus Wittich. London: University of California Press.
Werlen, Benno
  1993 Society, Action and Space: an Alternative Human Geography. London: Routledge.

Williams, Raymond
1973   The Country and the City. London: Chatto & Windus.
Williams, Rosalind
1990   Notes on the Underground: an Essay on Technology, Society, and the Imagination. London: MIT Press.
Williamson, J.
1982   Preface to the Fourth Impression. *In* Decoding Advertisements: Ideology and Meaning in Advertising, J. Williamson. London: Marion Boyars.
Wilson, D.
1994   Probably as Close as You Can Get to Paradise: Tourism and the Changing Images of Seychelles. *In* Tourism: the State of Art, A. V. Seaton, ed., pp. 765–774. Chichester: Wiley.
Wolff, K.
1950   The Sociology of Georg Simmel. Glencoe, Illinois: Free Press.
Wood, R. E.
1993   Tourism, Culture and the Sociology of Development. *In* Tourism in South-east Asia, M. Hitchcock, V. T. King and M. J. G. Parnwell, eds., pp. 48–70. London: Routledge.
Yiannakis, Andrew and Heather Gibson
1992   Roles Tourists Play. Annals of Tourism Research 19:287–303.
Young, G.
1973   Tourism: Blessing or Blight? Harmondsworth: Penguin.
Young, Michael and Tom Schuller, eds.
1988   The Rhythms of Society. London: Routledge.
Zacher, Christian K.
1976   Curiosity and Pilgrimage: the Literature of Discovery in Fourteenth-Century England. Baltimore and London: The Johns Hopkins University Press.
Zerubavel, Eviatar
1981   Hidden Rhythms: Schedules and Calendars in Social Life. London: University of California Press.
Zerubavel, Eviatar
1990   Private Time and Public Time. *In* The Sociology of Time, John Hassard, ed., pp. 168–177. London: Macmillan.
Zuckerman, M.
1979   Sensation Seeking: beyond the Optimal Level of Arousal. Hillsdale, NJ: Cawrence Erlbaum Associates.
Zukin, Sharon
1992   Postmodern Urban Landscapes: Mapping Culture and Power. *In* Modernity and Identity, Scott Lash and Jonathan Friedman, eds., pp. 221–247. Oxford: Blackwell.

Author Index

Subject Index